Lactose Hydrolysis in Dairy Products

Priyanka Singh Rao • Heena Sharma
Sumit Arora • Laxmana Naik N.
Editors

Lactose Hydrolysis in Dairy Products

Springer

Editors
Priyanka Singh Rao
Dairy Chemistry Section
Southern Regional Station, ICAR-National Dairy Research Institute
Bengaluru, Karnataka, India

Sumit Arora
Dairy Chemistry Division
ICAR-National Dairy Research institute
Karnal, Haryana, India

Heena Sharma
Dairy Technology Division
ICAR-National Dairy Research institute
Karnal, Haryana, India

Laxmana Naik N.
Dairy Chemistry Section
Southern Regional Station, ICAR-National Dairy Research Institute
Bengaluru, Karnataka, India

ISBN 978-3-031-78209-1 ISBN 978-3-031-78207-7 (eBook)
https://doi.org/10.1007/978-3-031-78207-7

This Springer imprint is published by the registered company Springer Nature Switzerland AG
The registered company address is: Gewerbestrasse 11, 6330 Cham, Switzerland

Preface

Lactose is the major sugar present in milk. It is a disaccharide made up of one unit of glucose and galactose, which is linked together by ß-1, 4 glycosidic linkages. Metabolism of lactose requires a specific enzyme: ß-galactosidase (lactase), which is present in the brush border of epithelial cells of the small intestine. Deficiency of this enzyme leads to malabsorption of lactose and fermentation of lactose occurs in gut, which results in various gastrointestinal disorders such as flatulence, intestinal discomfort, vomiting, and diarrhea. Around 70% of the world population suffers from lactose intolerance, out of which 90% population is Asian. In Indian scenario, approximately 75% of the population shows some signs of lactose intolerance. Such individuals face difficulty in consumption of milk and milk products which has augmented the demand of low-lactose or lactose-free milk products.

This book aims to consolidate current knowledge and advancements in the field of lactose hydrolyzed dairy products and foster innovation in lactose-free dairy product development amid increasing consumer demand. It also addresses the growing significance of lactose intolerance and the development of solutions to cater to affected populations. Beginning with the fundamentals of lactose intolerance followed by the biochemical basis and its prevalence across different demographics, it spans over the enzymatic hydrolysis of lactose as a key process in producing lactose-free dairy products.

As awareness of lactose intolerance grows, so does the demand for lactose-free options. Therefore, in this book, we attempted to bridge scientific understanding with practical application, providing insights into production techniques, health implications, and regulatory considerations of lactose hydrolyzed dairy products. Significance of lactose hydrolysis and challenges encountered during the manufacture of lactose hydrolyzed products along with opportunities within the food science landscape are highlighted. We strongly believe that this would serve as a useful resource for researchers, industry professionals, and the persons interested in the science, production, and market dynamics of lactose-free dairy products. I would like to express my gratitude to all the researchers, scholars, and experts in the field of food science whose contributions have enriched the content of this book. Their dedication to advancing knowledge and finding innovative solutions to lactose intolerance has been truly inspiring.

Bengaluru, Karnataka, India	Priyanka Singh Rao
Karnal, Haryana, India	Heena Sharma
Karnal, Haryana, India	Sumit Arora
Bengaluru, Karnataka, India	Laxmana Naik N.

Contents

Contributors

Abhishek Dutt. Tripathi Department of Dairy Science and Food Technology, Banaras Hindu University, Varanasi, Uttar Pradesh, India

Ananya Bharadwaj Department of Dairy Science and Food Technology, Banaras Hindu University, Varanasi, Uttar Pradesh, India

Sumit Arora Dairy Chemistry Division, ICAR-National Dairy Research Institute, Karnal, Haryana, India

B. Ashritha Dairy Technology Section, Southern Regional Station, ICAR-National Dairy Research Institute, Bengaluru, Karnataka, India

Deepak Dairy Technology Section, Southern Regional Station, ICAR-National Dairy Research Institute, Bengaluru, Karnataka, India

Avinash C. Gautam Dairy Technology Division, ICAR-National Dairy Research Institute, Karnal, Haryana, India

Nitika Goel Department of Dairy Technology, College of Dairy and Food Science Technology, GADVASU, Ludhiana, Punjab, India

Ankita Hooda Department of Dairy Science and Food Technology, Banaras Hindu University, Varanasi, Uttar Pradesh, India

Anusha Kishore Dairy Chemistry Division, ICAR-National Dairy Research Institute, Karnal, Haryana, India

C. T. Manoj Kumar Dairy Technology Section, Southern Regional Station, ICAR-National Dairy Research Institute, Bengaluru, Karnataka, India

Priyanka Meena ICAR-Central Sheep & Wool Research Institute, Avikanagar, Rajasthan, India

Laxmana Naik N. Dairy Chemistry Section, Southern Regional Station, ICAR-National Dairy Research Institute, Bengaluru, Karnataka, India

Pranali Nikam Department of Dairy Chemistry, College of Dairy Science and Food Technology, DSVCKV, Raipur, Chhattisgarh, India

Shreya Panwar Division of Crop Production, ICAR-Indian Institute of Vegetable Research, Varanasi, India

Sarang Pophaly Department of Dairy Microbiology, College of Dairy Science and Food Technology, DSVCKV, Raipur, Chhattisgarh, India

Nikheel Bhojraj Rathod PG Institute of Post-Harvest Technology and Management (Dr. Balasaheb Sawant Konkan Krishi Vidyapeeth), Raigad, Maharashtra, India

Mohammad Rizwan Dairy Microbiology Division, National Dairy Research Institute, Karnal, Haryana, India

Kanchanpally Saipriya Dairy Technology Division, ICAR-National Dairy Research Institute, Karnal, Haryana, India

M. H. Sathish Kumar Dairy Technology Section, Southern Regional Station, ICAR-National Dairy Research Institute, Bengaluru, Karnataka, India

Heena Sharma Dairy Technology Division, ICAR-National Dairy Research Institute, Karnal, Haryana, India

Rajan Sharma Dairy Chemistry Division, ICAR-National Dairy Research Institute, Karnal, Haryana, India

Sonu K. Shivanna Dairy Chemistry Division, ICAR-National Dairy Research Institute, Karnal, Haryana, India

G. P. Shivaswamy Agricultural Economics, Southern Regional Station, ICAR-National Dairy Research Institute, Bengaluru, Karnataka, India

Richa Singh Dairy Chemistry Division, ICAR-National Dairy Research Institute, Karnal, Haryana, India

Priyanka Singh Rao Dairy Chemistry Section, Southern Regional Station, ICAR-National Dairy Research Institute, Bengaluru, Karnataka, India

M. Sivaram Agricultural Economics, Southern Regional Station, ICAR-National Dairy Research Institute, Bengaluru, Karnataka, India

Arvind Soni ICAR-Central Sheep & Wool Research Institute, Avikanagar, Rajasthan, India

Sonia Department of Foods and Nutrition, CCSHAU, Hisar, India

S. Supreetha Dairy Technology Section, Southern Regional Station, ICAR-National Dairy Research Institute, Bengaluru, Karnataka, India

N. Veena Department of Food Safety and Quality, DSLD College of Horticultural Engineering and Food Technology, UHSB, Haver, Karnataka, India

Abbreviations

%	Percentage
AA	Amino acid
AGEs	Advanced glycation end products
ALU	One acid lactase unit
BCAA	Branched chain amino-acids
BIS	Bureau of Indian Standards
BSA	Bovine serum albumin
CAGR	Compounded annual growth rate
CLA	Conjugated linoleic acid
COS	Caprine oligosachharides
CSTR	Continuous stirred tank reactor
CVDs	Cardio-vascular diseases
DH	Degree of hydrolysis
DIAASs	Digestible indispensable amino acid score
dL	Deciliter
DNA	Deoxyribonucleic acid
DP	Degree of polymerization
E.C	Electronic configuration
ED	Electrodialysis
EFSA	The European Food Safety Authority
EMR	Enzymatic membrane reactor
ESD	Electrostatic spray drying
EU	European Regulations
FBR	Fluidized bed reactor
FCDB	Food Composition Data base
FD	Freeze drying
FOS	Fructooligosaccharides
FOSHU	Foods for specific health use
FSSAI	Food Safety and Standards Authority of India
g	gram
GCMMF	Gujarat Cooperative Milk Marketing Federation

GH	Glycoside hydrolase
GI	Gastrointestinal
GIT	Gastrointestinal tract
GOD	Glucose oxidase
GOI	Government of India
GOS	Galacto-oligosaccharides
GRAS	Generally Recognized as Safe
H	hours
HBT	Hydrogen breath test
HMF	Hydroxymethylfurfural
HPAEC	High-performance anion-exchange chromatography
HPLC	High-performance liquid chromatography
HPMCP	Hydroxypropyl methyl cellulose phthalate
Hz	Hertz
IAT	Internal air temperature
IDF	International Dairy Federation
Ig	Immunoglobulin
IHD	Ischemic heart Disease
IU	International units
kDa	kiloDalton
L	Liter
LB	*Lactobacillus delbrueckii* subsp. *bulgaricus*
LBA	Lactobionic acid
LC-MS	Liquid chromatography-mass spectrometry
LCT	Lactase
LDL	Low density lipoproteins
Lf	Lactoferrin
LFM	Lactose-free milk
LH	Lactose hydrolyzed
LHDMP	Lactose hydrolyzed dried milk powder
LHY	Lactose hydrolyzed yoghurt
LLPM	Low lactose powdered milk
LM	Lactose malabsorption
LOD	Limit of Detection
MBR	Membrane reactor
mg	Milligrams
Mg	Milligram
mL	Mililiters
MRPs	Maillard Reaction products
MSNF	Milk solid-not-fat
Mw	Molecular weight
NF	Nanofiltration
NIH	National Institute of Health
NLU	Neutral lactase units
OAT	Outlet air temperature

ONPG	*o*-Nitrophenyl-β-D-galactopyranoside
PAD	Pulsed amperometric detection
PBRs	Plug flow bioreactors
PDO	Protected designation of origin
ppm	Parts per million
QPS	Qualified presumption of safety
RDAs	Recommended Daily Allowance
RI	Refractive Index
rpm	Rotations per minute
SCFAs	Short chain fatty acids
SD	Spray drying
SMP	Skimmed milk powder
SRS-A	Slow reacting substance of anaphylaxis
ST	*Streptococcus thermophiles*
T_g	Glass transition temperature
TOS	Transgalactooligosaccharides
UDP	Uridine diphosphate
UF	Ultrafiltration
UHLM	Ultra-high temperature hydrolyzed-lactose milk
UHT	Ultra high temperature
USDA	United States Development Authority
USFDA	United States Food and Drug Administration
w/v	Weight/volume
w/w	Weight by weight
WPNI	Whey Protein Nitrogen Index
α-la	α-Lactalbumin
β-lg	β-Lactoglobulin

Chapter 1
Lactose Intolerance: An Overview

Pranali Nikam, Sarang Pophaly, and Mohammad Rizwan

1.1 Introduction

Milk and milk products are labelled as ideal balanced diet, as they contain many valuable nutrients for the human beings. Lactose plays significant role in the formation of glucocerebrosides; the vital components of nerve cells. Besides the essential nutrients, milk proteins, milk lipids, vitamins, and minerals, milk also contains lactose and other oligosaccharides which boost the progression of prebiotic bacteria (e.g., *Bifidobacterium*) in the small intestine of infants to safeguard them against the infections (Lambrini et al., 2021). Milk is a poor source of a few vitamins and iron, and still, it is considered as a perfect food for adults (Franzè & Bertelè, 2010). However, every individual cannot digest the milk, triggering in several gastrointestinal symptoms (Matthews et al., 2005; Harrington & Mayberry, 2008). Lactose is not only the source of calories; it also accelerates the absorption of several minerals such as calcium, phosphate, manganese and magnesium. While absorption of lactose, it should be first hydrolyzed in the human digestive tract by the substrate-specific membrane enzymes present in the intestine. The enzyme is lactase-phlorizin hydrolase (lactase). It acts on β (1 → 4) galactosidic linkage and splits the disaccharide molecule of lactose into the two monosaccharide components, i.e. glucose and galactose (Vandenplas, 2015). Some individuals are lacking the lactase enzyme

P. Nikam (✉)
Department of Dairy Chemistry, College of Dairy Science and Food Technology, DSVCKV, Raipur, Chhattisgarh, India

S. Pophaly
Department of Dairy Microbiology, College of Dairy Science and Food Technology, DSVCKV, Raipur, Chhattisgarh, India

M. Rizwan
Ph.D. Scholar, Dairy Microbiology Division, National Dairy Research Institute, Karnal, Haryana, India

P. S. Rao et al. (eds.), *Lactose Hydrolysis in Dairy Products*,
https://doi.org/10.1007/978-3-031-78207-7_1

activity in their gut and reported as lactose malabsorption. These individuals simply suffer with lactose intolerance (Vázquez et al., 2020). Lactose intolerance was first described by Hippocrates, (fifth century BC) and Galenus (second century AD). Since then, milk is known to be able to cause diarrhoea and other gastrointestinal manifestations in certain people.

Lactose intolerance is defined as the onset of gastrointestinal symptoms caused due to the reduction in lactase enzyme activity and leaving undigested lactose in the intestine. According to Epidemiological survey data, 70% of the world's population suffer from lactose malabsorption, which seriously impedes human absorption and utilization of milk nutrition because the lactase enzyme activity declines as the age increases (Li et al., 2023a, b). The gastro intestinal symptoms associated with this condition are commonly diarrhoea, abdominal cramping, nausea, vomiting, bloating, flatulence, etc. Lactose intolerance will also lead to deficiency of minerals including calcium and zinc in long term, which hinders growth of patients gradually. It has also been reported that these individuals may be at a risk of raising several other pathologies such as osteoporosis, irritable bowel disease, etc. Hence, proper diagnosis and treatment is necessary to deal with such conditions. Furthermore, the diagnosis of lactose intolerance is frequently affected by subjective judgement and milk intolerance.

1.2 Lactose Present in Milk and Dairy Products

Lactose is the principal sugar of mammalian milk, except *Pinnepedia* (sea lions and walruses) which may produce low volume, viscous and fatty-free milk. Lactose is a disaccharide comprising of D-galactose and D-glucose subunits, linked with (β-1,4) glycosidic bond and hydrolyzed in the small intestine by the enzyme lactase phlorizin hydrolase (colloquially, lactase) (Szilagyi & Ishayek, 2018). Milk is the only substantial source of lactose and its concentration varies from 0–10% (w/w). Lactose can be obtained from the whey (milk serum) via methods scuh as ultrafiltration, evaporation and subsequent crystallization and it was first isolated by the Italian scientist Fabrizio Bartoletti in 1633. It is six times less sweet than the sucrose. In milk, lactose has two isomers as: α-lactose and β-lactose (where, α and β represents the position of C_4 hydroxyl group of galactose). Both isomers differ in their chemical and physical properties including solubility, optical rotation, melting temperature, crystallization.

The concentration of lactose in milk is usually inversely proportional to its proteins and fats contents (Gänzle et al., 2008). The lactose is synthesized in epithelial mammary glands by the action of enzyme lactose synthetase from the two precursors as D-glucose and D-galactose which are absorbed from the blood, while in ruminants it is synthesized from the propionic acid. The enzyme lactose synthetase has two different subunits, as UDP-galactosyl-transferase (EC 2.4.1.22) activity and regulatory activity (α-lactalbumin). The subnunit UDP-galactosyl-transferase catalyze the transfer of galactosyl group from UDP-galactose to N-acetyl lactosamine and produce N-acetyllactosamine. Another is α-lactalbumin, which catalyzes the formation of disaccharide together with galactosyl transferase, and lactose from UDP-galactose and glucose (Brew, 2003).

The reported average concentration of lactose in mature cow and buffalo milk is about 4.8% (w/w) in each whereas goat and sheep milk have 4.1 and 4.6% (w/w), respectively. The lactose content in human milk was reported as 7.5%, (w/w). It was reported that the concentration of lactose during the early stages of lactation increased slightly but then at the end of lactation it showed decreasing trend to about 70% of the maximum. While, the concentration of lactose in the milk of *Equidae* (horse, donkey and zebra) increased during lactation, reaching values in the range 6.0–7.4%, w/w (Oftedal & Jenness, 1988; Doreau & Boulot, 1989). The average lactose content in milk, milk products and other foods is shown in the Table 1.1. Lactose powder is also used as food additive in many processed foods due to its physico-chemical properties such as textural and adhesive qualities, in addition to hydration and flavor improvements.

Table 1.1 Lactose content in milk, milk products and processed foods

Type of food	Lactose content in g/100 g of food	Lactose content/serving in g
Milk		
Cow milk	4.8	15
Skimmed milk	4.8	15
Buffalo milk	4.8	15
Goat milk	4.5	13
Sheep milk	4.6	14
Buttermilk	3.0	9.0
Lactose-free milk	<0.1	<0.1
Milk products		
Biological yoghurt	4.0	9.5
Fresh yoghurt	3.0	9.3
Cream	3.6	3.2
Butter	0.5	0.1
Cream Cheese	3.0	0.9
Hard Cheese	0.1	<0.1
Soft Cheese	0.3	0.1
Soft ice cream	6.4	5.7
Other food stuffs		
Pudding/custard	3.6	4.5
Cheese burger	0.9	1.1
Lasagne	1.1	2.6
Ready sauces	3.6	4.5
Latte Macchiato	4.3	8.6

Compiled from Pansuwan (2023), Misselwitz et al. (2019)

1.3 Lactose Malabsorption

Lactose malabsorption is considered as a physiological problem when sizable fraction of ingested lactose is non-digested or not absorbed at the brush border of small intestine. It is mainly because of the low lactase activity (Kaur et al., 2006). There are two types of conditions based on degree of lactase activity. One is alactasia which indicates the total absence of lactase activity and other is hypolactasia which indicates the deficiency of lactase activity in the jejunal mucosa hence, the ingested lactose cannot be hydrolyzed into an absorbable component in the intestine (Misselwitz et al., 2019). The lactose metabolism is highly dependent upon the activity and expression of the lactase enzyme. The various terms related to the lactase deficiency, lactose malabsorption and lactose intolerance is defined in the Table 1.2.

1.3.1 Lactose Digestion

The digestion of lactose takes place in the small intestine of human body (Fig. 1.1). Hence, to achieve the metabolic role of lactose, it should be hydrolyzed by lactase-phlorizin hydrolase (LPH) [3.2.1.23] enzyme available in brush border villi with its abundant activity occuring in mid-jejunum layering with the epithelium of fully matured enterocytes. It consists of two identical extracellular polypeptide chains (Mw 160 kDa) with α-glucosidase and β-galactosidase activity (Misselwitz et al., 2019). This enzyme acts on specific substrate lactose and cleaves in two monosaccharides units i.e., D-glucose and D-galactose. The lactase enzyme shows its optimal activity at pH 6.0–8.0 in the small intestine. Basically, lactase enzyme has two active sites; one catalyzes the hydrolysis of lactose, and second breaks phlorizin (an aryl α-glucoside). The membrane proteins then absorb the resultant monosaccharide sugars and enters into the enterocyte via sodium-glucose linked transporter 1 (SGLT1), with the simultaneously movement of two Na^+ ions from the intestinal lumen via active transport mechanism. The monosaccharides then enter into the bloodstream either by facilitating protein GLUT 2 (glucose transporter 2) or by passive transport. Glucose is then utilized as an energy source, whereas galactose is used as a constituent of glycolipids and glycoproteins or converted into glucose using galactokinase and galactose 1-Pi-uridyl transferase.

Due to the lactase deficiency in the intestine, the non-digested/non-absorbed lactose has high osmotic load which pulls high amount of intestinal fluid and electrolytes. The colonic microbiota easily ferments the non-digested lactose and leads to production of various gases mainly carbon-dioxide (CO_2), hydrogen (H_2) (Fig. 1.1). These gases are further utilized by the methanogenic bacteria and releases methane (CH_4) gas which may further enter to the bloodstream. These gases are eliminated from the body through breathing. These gases also cause bloating effect. The synthesis of short-chain organic acids e.g., propionic acid, succinic acid, formic acid,

Table 1.2 Terms related to the lactase deficiency, lactose malabsorption and lactose intolerance

Concept	Explanation	References
Congenital lactase deficiency (CLD)	It is a rare genetic disorder (frameshift mutation) leading to an inability to produce lactase enzyme. It shows symptoms immediately after the birth.	Pansuwan (2023) and Ugidos-Rodríguez et al. (2018)
Neonatal lactase deficiency or Developmental lactase deficiency	It is observed in premature infants commonly <34 weeks, born before lactase begins to be expressed in the epithelium. It vanishes when the intestinal mucosa matures.	Shafi and Husain (2022) and Ugidos-Rodríguez et al. (2018)
Adult type hypolactasia or Primary lactase deficiency	It is the prevalent cause of lactase deficiency. It is due to changing in polymorphisms in the transcription promoter region of β-galactosidase gene.	Shafi and Husain (2022) andUgidos-Rodríguez et al. (2018)
Secondary lactase deficiency	It is caused due to small intestinal mucosal injury, such as persistent diarrhea, acute gastroenteritis, small bowel bacterial overgrowth, and cancer chemotherapy.	Shafi and Husain (2022) and Ugidos-Rodríguez et al. (2018)
Lactase non-persistence	Reduction in the intestinal lactase activity in first 20 years of life. This phenotype is found worldwide.	Pansuwan (2023)
Lactase persistence	It is the continuous expression of intestinal lactase enzyme beyond infancy, the phenotype commonly found in Western counties	Pansuwan (2023) and Misselwitz et al. (2019)
Lactase deficiency	Non-digestion of large amounts of lactose due to low activity of lactase enzyme in the small intestine.	Pansuwan (2023)
Lactose malabsorption	It is the passage of lactose into the colon doe to lactase deficiency or other pathologies.	Pansuwan (2023) and Misselwitz et al. (2019)
Primary lactose malabsorption	It is due to lactase non-persistence (common phenotype observed worldwide).	Pansuwan (2023) and Misselwitz et al. (2019)
Secondary lactose malabsorption	It is caused due to low lactase activity, typically caused by acute enteritis. It can be reversible.	Pansuwan (2023) and Misselwitz et al. (2019)
Lactose intolerance	It is the digestive disorder causes symptoms such as abdominal pain, diarrhea, bloating in persons with lactose malabsorption	Pansuwan (2023) and Misselwitz et al. (2019)
Functional lactose intolerance	Symptoms of lactose intolerance observed in persons without lactose malabsorption	Pansuwan (2023) and Misselwitz et al. (2019)
Self-reported lactose intolerance SLI.	Symptoms of lactose intolerance based on the past history without formal testing of lactose intolerance/lactose malabsorption	Pansuwan (2023) and Misselwitz et al. (2019)

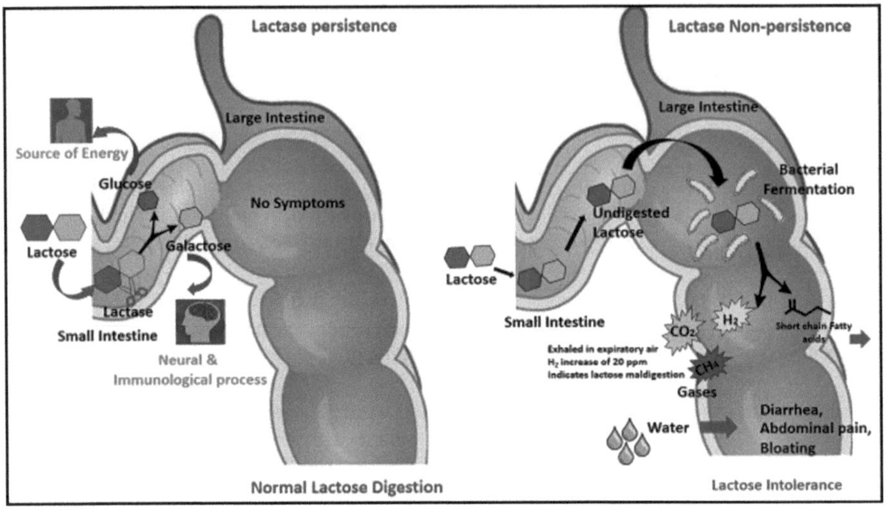

Fig. 1.1 Schematic representation of normal lactose digestion and lactose intolerance

lactic acid, and butyric acid during the fermentation causes the lowering of the colonic pH and hence the gastrointestinal function is affected (Lomer et al., 2008; Amiri et al., 2015).

1.3.2 *Lactose Intolerance*

Lactose intolerance is the onset of gastrointestinal disorder due to the lactose malabsorption which triggers the symptoms such as bloating, flatulence, abdominal cramps, nausea, diarrhea, vomiting or there may be more than usual burping. The signs and symptoms usually appears in 30 min to 1–2 h after the consumption of lactose containing foods, such as dairy products. The lactase deficiency causes the non-digestion of ingested lactose due to incomplete hydrolysis and creates lactose maldigestion. The resulting GI symptoms are due to the fermentation of undigested lactose. The types of lactose intolerance are explained in Table 1.2.

Although, the severity of lactose intolerance depends on the intrinsic as well as extrinsic factors such as expression and activity of lactase enzyme, lactose dose in the diet, intestinal flora and gastrointestinal tract for absorption of lactose and other fermentation products of lactose malabsorption (Catanzaro et al., 2020). There is no threshold level of lactose dose for causing lactose intolerance in individuals. Study revealed that the consumption of 50 g of lactose leads to symptoms in 80–100% of patients who are hypolactasic, however those who are extremely intolerant may create symptoms in just 3 g of lactose consumption ($\approx$ 60–70 mL cow milk). The majority of individuals can tolerate lactose up to 10–12 g ($\approx$ 200–250 mL milk) in a day with no substantial symptoms (Deng et al., 2015).

1.3.3 Lactose Intolerance and Milk Protein Allergy

Milk produces two types of adverse effects after ingestion, one is lactose intolerance and other is milk allergy. Lactose intolerance is basically non-immunological disorder, caused due to inability of digestion of milk sugar i.e., lactose due to the deficiency of lactase enzyme. For the lactose intolerance, lactose malabsorption is an essential precondition. While, milk allergy is termed as immunological disorder caused by the partially or complete degradation of milk proteins. The most allergic proteins are α_{S1} casein, α_{S2} casein and β-lactoglobulin. Among these two adverse reactions, the most prevalent adverse effect is lactose intolerance, mostly common amid the youngsters and adults, after milk allergy. The duration of milk allergy is very short and limited among the infants and kids. The milk protein composition and allergens are reported in Table 1.3.

Table 1.3 Bovine milk protein composition and allergens

Protein (30–35 g/l)		MW (kDa)	Composition (mg/mL)	No. of AA residues/ molecules	No. of S-S bridges/ molecules	Main functions	Allergens
Caseins (80%)	α_{s1} casein (32%)	23.6	11.6	199	0	Calcium binding	Bos d 9 (Major)
	α_{s2} casein (10%)	25.2	3	207	1	Calcium binding	Bos d 10 (Major)
	β casein (28%)	24.0	9.6	209	0	Calcium binding	Bos d 11 (Major)
	κ casein (10%)	19.0	3.6	169	1	Milk stability	Bos d 12 (Major)
Whey proteins (20%)	β-lg (10%)	18.3	3.0	162	2 +1 free -SH	Lipid and retinol binding protein	Bos d 5 (Major)
	α-la (5%)	14.2	1.2	123	4	Calcium binding lactose synthesis	Bos d 4 (Major)
	Igs (3%)	<150	0.6	–	Depend on class and subclass	Provides immunity	Bos d 7 (Minor)
	BSA (1%)	66.4	0.4	582	17 +1 free -SH	Ligands transport and protect from free radicals	Bos d 6 (Minor)
	Lf (Traces)	76.2	0.3	703	16	Iron binding protein and antimicrobial activity	— (Minor)

[a]MW- Molecular weight, AA- Amino acid
(Source: Fox et al., 2015, Bøgh & Larsen, 2021; Asrar et al., 2023)

Milk proteins allergy is a complex kind of food allergy and also the misapprehension chaos between the general public health due to revelation amongst the lactose intolerance and milk protein allergy. For both disorders, a distinct technique is required to diagnose and approaches to treat and control. Milk proteins allergy is abnormal condition arise due to the production of large number of antibodies rather than the specific antibodies (immunoglobulin) in respond to our body's immune system. These allergic symptoms able to harm body tissues due to deviant inflammatory respond. Allergic symptoms are similar to the antigen-antibody reactions, but it is aberrant to some of the individuals who are subtle to allergens. Milk proteins allergy consisted of continuum immunological mechanisms, usually distinguish as IgE mediated and Non-IgE mediated allergy. In IgE mediated milk proteins allergy, the inception of symptoms is rapid, occurring within minutes to an hour after allergen exposure, IgE-mediated allergy is often referred to as "immediate hypersensitivity". The IgE has specificity to bind on surface of certain cells, basophils and mast cells. These cells are present in large amount over the respiratory linings and GI tracts and also below the skin. IgE can then stimulates the production of substances which are responsible for allergic reactions like histamine, kinins, anaphylaxis (SRS-A) and eosinophil chemotactic factors (ECF-A) (Walsh et al., 2016). Non-IgE-mediated reactions tend to be delayed, with the onset of symptoms occurring from 1 h to several days after ingestion of milk. Hence, they are often referred to as "delayed hypersensitivity" (Sabra et al., 2003). The IgE mediated cow milk allergy is more dominant in children about 2–6% and quite low in adults (0.1–0.5%) (Sicherer & Sampson, 2018; Arasi et al., 2022). The common symptoms may include diarrhea, vomiting, skin rashes, difficulty in breathing (Walsh et al., 2016). The most common observed cases were skin manifestation (5–90%), anaphylaxis (0.8–9%) and GIT symptoms (60%) (Connor et al., 2022).

1.3.4 Diagnosis of Lactose Malabsorption

The degree of lactose malabsorption can be assessed by proper diagnosis. It can be diagnosed by taking several histories such as eating history, nutritional history, clinical history and physical examination etc., but the patients whose history taking physical examination can't be clearly diagnosed. Hence, the additional diagnostic tests are needed for confirmation. Two types of diagnostic tests are available as indirect and direct diagnosis.

1.3.4.1 Indirect Diagnosis

(a) **Provocation test**

In this test, the lactose containing products, such as milk and milk products are omitted from the daily diet for the period of **15** days, and then reintroduced. As the symptoms will vanish and then return as it is reintroduced at the end of the period, it can be suggested that the person is lactose intolerant.

(b) **Hydrogen breathe test**

This test is simple, non-invasive and most commonly used for diagnosing the lactose intolerance. It measures the excretion of H2 gas in respiratory tract after the oral consumption of lactose as a standard dose (**20–25** g of lactose in water) (Robles & Priefer, **2020**; Schnedl et al., **2020**). As the H**2** gas cannot be produced by the mammalian enzymes, so it indicates the interaction of carbohydrates with intestinal bacteria and reported as lactose malabsorption. The H**2** gas is circulated through the intestine and then passes in to blood. It then reaches to pulmonary alveoli, and is expelled in the breath. This test is positive as level of H**2** gas is greater than **20** ppm in **3** h of the exhaled air. False-negative results may also appear due to non-H**2** producing microbiota. Methanogenic bacteria (e.g, ***Methanobrevibacter smithii***) in some patients converts H2 into CH**4** in **4**:**1** ratio, gives a lower H**2** gas and results in positive test (Rezaie et al., **2017**; Varju et al., **2020**). Therefore, the use of **13**C lactose with breath measurements of **13**CO**2** is considered as a marker of lactose digestion and H2 as a marker of lactose metabolism can be opted (Houben et al., **2015**).

(c) **Stool acidity test**

The stool acidity test measures the pH of stools. The undigested lactose which is then fermented by intestinal bacteria producing the lactic acid, volatile fatty acids may lower the stool pH. This test is considered as non-specific and has very low sensitivity. The pH <**5.5** in stool indicates the lactose malabsorption.

(d) **Lactose tolerance test**

In this test, the level of glucose is measured in blood plasma after the ingestion of **50** g of lactose at different intervals of time e.g., **0**, **30**, **60**, **120** min. Lactose intolerance test is positive if glucose level is more than **20** mg/dL. But in this test, false positive results may occur in about **30**% cases because of quick insulin response (Domínguez Jiménez & Fernández Suárez, **2016**).

(e) **Gaxilose test or Lactest**

In this test, oral administration of gaxilose (**4**-galactosyl xylose, lactase substrate) is given and the amount of D-xylose is measured in urine samples or blood after **24** h. This test is simple, low in cost, sensitive and specific with the accuracy of **93**% which is higher than the H**2** test (**85**%) and lactose tolerance test (**79**%) as compared to duodenal biopsy (Aragón et al., **2014**).

1.3.4.2 Direct Diagnosis

(a) **Genotypical studies**

In this test, Real-Time Polymerase Chain reaction (RT-PCR) is used for the sequencing of the genomic DNA which can be extracted from the blood samples or buccal swab samples. This method identifies various polymorphisms of MCM**6** gene phenotype and marks as intolerant. This test has higher specificity (**100**%) and sensitivity (**93%**). This test is considered as quick but has disadvantages such as high cost and capital intensive (Enattah et al., **2002**).

(b) **Lactase enzyme activity test (Intestinal biopsy test):**

This test is based on the principle of analyzing the activity of lactase enzyme on intestinal (duodenal) biopsy. This test detects both primary and secondary lactose malabsorption. The assay is based on colorimetric reaction between the tissue samples obtained from duodenum and glucose oxidase which liberates glucose from the lactose after the incubation for **20** min (Ugidos-Rodríguez et al., **2018**). This test effectively detects the severe duodenal lactose malabsorption with **100%** specificity and **95%** sensitivity (Mattar et al., **2012**). However, this test requires specialized laboratories due to the biopsy protocol.

1.4 Health Issues Due to Lactose Intolerance

1.4.1 Bone Health

Among all the benefits traditionally known to be associated with consumption of dairy products, bone health is considered as one of the most important aspects (Rizzoli, 2022). In popular perception, this health benefit of dairy products is often associated and limited to presence of high calcium content, but in the arena of nutrition, presence of other milk constituents are now known to play a significant role along with calcium (Palacios, 2006; Ilich & Kerstetter, 2000). Bone formation requires copious amounts of calcium along with protein, magnesium, phosphorus, vitamin-D, potassium and fluoride. Metabolic processes related to bone require additional nutrients such as manganese, copper, boron, zinc, vitamin A, vitamin K, vitamin C and certain B vitamins (Palacios, 2006). There is overwhelming evidence in favor of dairy products supporting bone health by improving body bone mineral content (BMC), increasing bone mineral density (BMD) and reduction in fracture risk and prevention of osteoporosis in various stages of human life (Caroli et al., 2011; Heuvel and van den Heuvel & Steijns, 2018; Kun et al., 2001; Chan et al., 1995; Infante & Tormo, 2000).

Lactose intolerance, on its face, seems to be a simple inability to metabolize dietary lactose. The direct effects associated with lactose intolerance are therefore often related to gut problems. Since individuals with severe lactose intolerance have to avoid consumption of dairy products, they may indirectly be affected by deficiency diseases arising as a result of exclusion of milk-based products (Kudlacek et al., 2002). It is common among individuals to consider self as lactose intolerant even without undergoing any clinical diagnostics. Such self-perceptional or even actual lactose intolerance, both lead to considerations in poor life-quality (Casellas et al., 2016). Avoiding dairy products lead to inadequate calcium and vitamin D intake, which predisposes these individuals to decreased bone accrual, osteoporosis and other adverse health outcomes (Suchy et al., 2010).

Lumbar and femoral BMD (bone mineral density) was found significantly different in lactose tolerant and intolerant subjects preventing achievement of adequate peak bone mass and predisposing the latter group to severe osteoporosis (Di Stefano et al., 2002).

At genetic level, LCT (T/C-13910) polymorphism is linked with lactose intolerance in adults. Genetic lactase deficiency in women having the CC genotype was linked to significant bone fracture incidence which is in-turn linked with lower milk calcium intake and aversion to milk consumption (Obermayer-Pietsch et al., 2004). Interestingly, the CC genotype is also associated with a lower plasma vitamin D concentrations (Alharbi & El-Sohemy, 2017). Dietary phosphate is important for cartilage and osteoid tissue mineralization (Murshed, 2018) and impaired growth and fractures are commonly associated with phosphate wasting syndromes. Reduced consumption of dairy products can be linked to phosphate deficiency and a consequent lowered bone health (Rizzoli, 2022).

1.4.2 Cardiac Health

Cardiovascular disease (CVD) remains a significant global health concern. Cardiovascular diseases, including heart disease and stroke, continue to be the leading causes of death worldwide. They account for a significant proportion of deaths in both developed and developing countries. Cardiovascular diseases are the leading cause of death globally. According to the World Health Organization (WHO), an estimated 17.9 million people died from CVDs in 2019, representing 32% of all global deaths (Roth et al., 2020).

Several modifiable and non-modifiable risk factors contribute to the development of CVDs. Modifiable risk factors include unhealthy diet, physical inactivity, tobacco use, harmful use of alcohol, obesity, high blood pressure, high cholesterol, and diabetes. Non-modifiable risk factors include age, gender, and family history. Cardiovascular diseases impose a significant economic burden on healthcare systems and societies due to healthcare costs, loss of productivity, and disability. Effective prevention, early detection, and management of CVDs are essential to mitigate this impact.

Prevalence of CVDs has been associated both positively and negatively with the consumption of dairy products. Extensive epidemiological as well as experimental data is available on the contribution or involvement of dairy products in initiation or progression of CVDs. But there is a scarcity of data on presence of lactose or inability to digest lactose with respect to prevalence of CVDs. One factor which is broadly considered is the exclusion of dairy products as a result of lactose intolerance from the diet. This leads to a progressive absence of certain nutrients from the diet which have a direct impact on hearth health. Epidemiological data, though inconclusive, in certain studies have shown, that presence of lactose or inability of subjects to digest the disaccharide, does have specific affects with respect to CVDs particularly Ischaemic heart disease (IHD) (Segall, 1994). Studies on ethnic data have shown that a diet low or relatively low in lactose, in populations with low or relatively low prevalence of lactose absorbers, is more consistently associated with protection against ischaemic heart disease than are high intakes of unsaturated fatty acids, wine, alcohol or dietary fibre. In seven countries with a high consumption of dairy products (six at least with a high prevalence of lactose absorbers), trends in ischaemic heart disease mortality appear to have reflected changes in the supply of milk (and therefore of lactose), but not consistently of butter or inversely of unsaturated fatty acids (Segall, 1994). In a similar cross analysis study, the prevalence of primary adult lactose malabsorption (LM) in 23 ethnic groups was matched with national data on milk consumption and mortality rates from IHD. The study revealed that, in comparison with populations who have a low prevalence of LM and a high milk consumption, populations with high and intermediate prevalence of LM have a low risk of IHD mortality if their consumption of milk is low or is largely in low lactose form (Segall, 2008).

Another aspect which was gaining traction to certain extent is the use of lactose in pharmaceutical drugs which might trigger certain symptoms in sensitive individuals. Lactose is often used as an excipient in pharmaceutical drugs. Certain studies indicated that the amount of lactose in most drugs is not sufficient to cause symptoms in persons with lactose intolerance, although inter individual differences in sensitivity probably exist (Vinther et al., 2015). Many CVDs drugs use lactose as an ingredient. In one case study, an individual with severe lactose intolerance was reported to undergo treatment of ischemic cardiomyopathy. The patient was discharged on dual antiplatelet therapy, rosuvastatin, perindopril and carvedilol, he was repeatedly readmitted in the following days for abdominal pain/bloating, diarrhea and nausea despite avoiding food products containing lactose (Craveiro Barra et al., 2012).

1.4.3 Colorectal Cancer and Inflammatory Bowel Diseases

Colorectal cancer is the third most common cancer worldwide. Colorectal cancer has multifactorial aetiology and can be attributed to various predisposing conditions and diet-related issues. More than any other form of cancer, the incidence and progression of colorectal cancer is closely linked to diet. Dairy foods have long been considered as having beneficial effects with respect to colon cancer. There has been an empirical knowledge of dairy products being beneficial and/or having a preventative effect in this type of cancer (Pophaly et al., 2022). Milk constituents such as vitamin D, calcium, CLA, sphingolipids have been shown to have a protective effect against colon cancer (Kim, 2017). Fermentation of milk constituents by intestinal bacteria causes release of SCFA such as butyric acid, which is crucial for epithelial stability in the gut. Calcium and vitamin D regulate cell growth and promote cell differentiation by stimulating calcium-sensing receptors, and calcium itself forms nontoxic insoluble compounds by binding with vulnerable free fatty acid and free bile acids (Vargas & Thompson, 2012).

Most of the correlations between the consumption of dairy products and perceived protective or therapeutic effects have been derived from cohort and meta-analyses studies in different population groups. The exclusion of dairy products from diet and a possible consequent prevalence of colon cancer in population, thus was intriguing. Epidemiological studies in Hungarian (Bácsi et al., 2008), Finnish (Rasinperä et al., 2005) and Polish populations (Andrzej et al., 2015) have shown increased risk of colorectal cancer among lactose intolerant people. But these reports have not been consistent with contradictory results showing no relation between two phenomenon's in British, Hungarian, Spanish and Italian population. (Tarabra et al., 2010). These epidemiological studies need to be backed up with strong cause and effect trials and experimental data. The debate and evidence on prevalence of cancer in correlation to lactose intolerance, does not deny the fact that lack of a dairy based diet will cause severe deficiency of certain nutrients which are known to influence the prevention and treatment of the disease. The most documented and widely accepted effect has been observed for milk calcium and vitamin D, when consumed both alone or in combination. Increased serum levels of vitamin D and calcium have been associated with lower risks of colorectal cancer (CRC) incidence and mortality (Zhu et al., 2017). These inverse associations may be mediated by the vitamin D receptor (VDR) and the calcium-sensing receptor (CASR) (Zhu et al., 2017; Yang et al., 2018).

1.5 Treatment Options for Lactose Intolerant Population

1.5.1 Alternative Dairy Products for Lactose Intolerant Population

Lactose intolerance had led to a considerable population excluding dairy products from their dietary habits. This in-turn has been linked to malnourishment and associated disorders. To put dairy products back on the platter of lactose intolerance people, major scientific and technological interventions were undertaken by academia and dairy industry. The challenge was to remove lactose from milk and milk products, while keeping other ingredients and nutrients intact as far as possible. This was primarily achieved by introduction of neutral exogenous lactase (β-galactosidase) which can predigest lactose (Nakayama & Amachi, 2002). Commercial production of this enzyme is done by limited companies around the world like DSM Food Specialties, (the Netherlands) and Godo, Nagase and Amano carry out the operations in Japan (Dekker et al., 2019). These enzymes are available through select channels marketed by more famous brand names such as Halactase (Chr. Hansen), LactozymePure (Novoeymes), Godo YNL2 (Dupont). The lactose free dairy market was approximately USD 10.7 billion in 2021 and is estimated to reach a valuation of USD 17.4 billion by 2026 at a CAGR of 5.4%.

The commercial manufacturing of lactose free milk (LFM) was undertaken in 1970s. The enzymatic process for lactose hydrolysis does not require extensive equipment in dairy plants. Single enzyme treatment for hydrolysis should be based on several factors such as substrate concentration, pH, and maximum temperature employed during processing, contact time and of course the cost of enzyme treatment. An extended contact time at temperature range of 35-45 °C may give sufficient hydrolysis but can also result in significant microbial growth (Zadow, 1986). Microbial lactases from *Aspergillus niger*, *Aspergillus oryzae*, *Kluveromyces lactis*, *Escherichia coli*, *Bacillus circulans*, *Bacillus subtilis*, *Lactobacillus acidophilus* and *Streptococcus thermophilus* are commonly used (Harju et al., 2012). Hydrolysis of lactose and concomitant production of constituent monosaccharides increases the sweetness of the product. If the product involves sucrose addition, then appropriate changes in sucrose levels should be followed so as to balance the overall sweetness profile. In heat treated products, the high levels of individual monosaccharides react at an enhanced rate with lysine and release stable Maillard reactions products (MRPs) (Singh et al., 2021).

Acid hydrolysis, as compared to enzyme-based processes is technologically challenging and more cost intensive and hence not widely practiced in dairy industry. Protein free solutions such as UF permeate of milk and whey can be delactosed using acid hydrolysis. The pH of the stream is adjusted to 1.2 and temperature to 150 °C for a short duration. The hydrolyzed product is sensorily inferior and requires extensive treatments such as neutralization, demineralization and decolorization. Membrane based methods are also not very popular because of high initial cost and problems of microbial growth. Protein free stream such as milk or whey UF permeate is fed and enzyme is recovered from the reaction mixture with a second UF equipment and the permeate containing hydrolyzed lactose (sans enzyme) is reintroduced with milk or whey retentate (Harju et al., 2012).

For long, the nutritional requirement of lactose intolerant people was not dealt at a policy or state level. The lack of awareness and low volume of the market was another reason to it. But in recent years, policymakers, academia and industry have shown a trend towards low or lactose free products. A very unique initiative in this direction was carried out, with the development of a database of lactose free products commercially sold in Spain called as "Lactose free food composition database (FCDB)" (Martínez Rodríguez et al., 2021). The aim of this database is to evaluate the nutritional composition of lactose-free products from Spain. The database is compiled from market survey of all the products with "lactose-free" label on them. The database counts for 327 products out of which 123 were dairy products, 16 are food products with dairy as an ingredient and 188 are non-dairy products with "lactose free" labelling on them. The main subgroups are yogurt (25%), milk (24%) and cheese (17%) (Martínez Rodríguez et al., 2021). Such market survey databases should be developed in other countries especially where lactose intolerance is more prevalent. This will give a better idea of per capita availability of these products in a market and also the food choices available for consumers. A list of some lactose free dairy products and their outcome is listed in Table 1.4.

Table 1.4 Lactose free dairy products

S. no.	Product	Intervention/Outcome	References
1	Lactose free milk	Graphene oxide membranes used for lactose separation	Morelos-Gomez et al. (2021)
2	Lactose free milk	Low temperature hydrolysis of lactose by commercial enzyme NOLA™ Fit 5500	Czyzewska and Trusek (2021)
3	Lactose free probiotic goat milk beverage	Supplementation of jambo pulp and addition of probiotic culture	Araújo et al. (2022)
4	Lactose hydrolyzed whey	β-gal from *Aspergillus niger* and *Aspergillus oryazae*	Nakayama and Amachi (2002)
5	Lactose free yogurt	Lactase enzyme (brand: Lactazik) used for hydrolysis; Standard yogurt culture used for fermentation	Capcanari et al. (2021)
6	Low lactose milk powder	Combination of three membrane technologies (UF, NF and ED) for physical removal of lactose	Zhang et al. (2020)
7	Lactose free functional yogurt	Lactose hydrolyzed with β-gal and supplemented with Fructooligosaccharide (FOS)	Pachekrepapol et al. (2021)
8	Low lactose yogurt	Growth of *E. coli* O157:H7 was studied in lactose free milk	Cutrim et al. (2016)
9	Lactose free frozen yogurt	Addition of k-carrageenan and corn starch as stabilizer in frozen yogurt	Skryplonek et al. (2019)
10	Lactose free frozen yogurt	Commercial Ha-lactase® was used for hydrolysis which improved the texture and viscosity of product and enhanced sensory qualities.	Skryplonek et al. (2017)
11	Lactose free yogurt	Incorporation of uvaia pulp	Bianchini et al. (2020)
12	Lactose free ice cream	Lactose hydrolysis causes improvement on textural properties of ice cream	Abbasi and Saeedabadian (2015)
13	Lactose free cheese	Residual lactose removed through washing curd with hot water	da Silva et al. (2020)
14	Lactose free mozzarella cheese	Sensory profile and shelf life of lactose free mozzarella was compared with traditional mozzarella cheese	Cincotta et al. (2021)
15	Lactose free ice cream	Scuralose was a better sweetener for lactose free ice cream as compared to stevia	de Medeiros et al. (2021)

1.5.1.1 Lactose Free Fermented Milks and Cheeses

Fermentation is a natural way to hydrolyze the lactose in milk (Solomons, 2002). Fermented milk products do have lower lactose content as partial hydrolysis converts it to lactic acid. Still a substantial amount of lactose is present in the final fermented product (Dekker et al., 2019). Though, the trigger point in terms of lactose concentration for intolerant individuals can vary, the low or trace lactose still can cause problems in certain cases. Therefore, hydrolysis of residual lactose in fermented milk products needs to be achieved. It can be done by either incubating the milk with lactase before heat treatment or by adding lactase together with the culture after the pasteurization process. The latter approach of co-hydrolysis is more widely adopted. Predigestion may cause inhibition or slow growth of starter culture and hence an undesirable product. In co-hydrolysis method, the dosage of enzyme should be higher as there is gradual decrease in the enzyme efficacy due to pH drop as the fermentation proceeds. Most neutral lactases are inactive below the pH of 5.5, which is normally achieved in 3 h of incubation. Acid stable lactases are also available in the market which has a longer residual hydrolysis activity.

The functional value of lactose free fermented products is enhanced by incorporating or fortifying with additional nutrients. Lactose free Yogurt supplemented with chia seeds, flax seeds, sesame seeds and yogurt fortified with iron was evaluated for physicochemical and technological aspects (Capcanari et al., 2021). Bianchini et al. (2020) tested the addition of uvaia pulp in lactose free yogurt, which increase the phenolic content, total carotenoids, and antioxidant activity of the final yogurt. Probiotic lactose free yogurt was prepared by addition of probiotic culture *Bifidobacterium* BB-12 in encapsulated form, to lactose free yogurt to develop a Skyr style product. Ayran is salty drinkable yogurt. Sensory properties of Ayran produced from 50% and 100% lactose hydrolyzed milk was significantly different from commercial Ayran. Syneresis and flow behavior index values increased, and viscosity values decreased with increasing lactose hydrolysis level, salt content and storage period (Kocabaş et al., 2022).

Cheese is a dairy product which is mostly devoid of lactose, as the disaccharide is mostly drained along with whey. When rennet induced coagulation of milk is achieved, the curd is precipitated out and the whey portion is drained. The soluble lactose is dissolved in the whey and thus the cheese mass is practically free of lactose. Although residual lactose is always present and its concentration may vary from variety to variety and also with respect to certain processing conditions. The products which are devoid of lactose due to separation during manufacturing are categorized as "naturally lactose free". Twenty five cheese varieties (with Protected Designation of Origin) were screened for residual lactose content using LC/MS-MS methodology with a limit of detection of 10 mg/kg (Facioni et al., 2021). The study showed that majority of cheese varieties were naturally lactose free. The extent of lactose removal was influenced by factors such as composition of starter culture, extent of ageing and environmental factors. Grana Padano, a traditional hard cheese variety, due to its extremely low lactose concentration as well as lower glucose and galactose we well, is considered as one of the most suitable cheese for lactose intolerant people (Monti et al., 2017).

1.5.2 Alternative Non-Dairy Products for Lactose Intolerant Population

In simple sense, all the non-dairy products of the world can be considered as an option for lactose intolerant people. But the products which can replace bovine milk in terms of its technological and sensory requirements, are more suitable in the category of non-dairy products for lactose intolerant. These can include plant-based milk and derived products from such milks (Table 1.5).

Table 1.5 Non-dairy products for lactose intolerant individuals

S.No,	Product	Functionality	References
1	Oat milk	Supports healthy cholesterol; rich in vitamins and fibre	Cui et al. (2023)
2	Almond milk	Improvement of vision, weight control, stronger bones, less calories	Kundu et al. (2018)
3	Soy milk	Rich in proteins, balanced diet	Vanga and Raghavan (2018)
4	Coconut milk	Low in calories, taste is favored	Vanga and Raghavan (2018)
5	Rice milk	Comparable calories to cow milk	Vanga and Raghavan (2018)
6	Hazelnut milk	High antioxidative activity, high nutritive value	Maleki et al. (2015) and Aysu et al. (2022)
7	Walnut milk	High concentration of mono and poly unsaturated fatty acids	Popovici et al. (2017)
8	Tofu	Potential replacement of *Paneer* or cottage cheese	Parmar et al. (1989)

1.5.3 Oral Enzymes

Orally administered enzymes are currently in clinical use or under development for a range of pathologies. The oral route for delivery of enzymes is challenging due to inherent sensitivity of these proteins to harsh conditions in the gastrointestinal tract (Fuhrmann & Leroux, 2014). Exogenous oral lactase in form of chewable tablets or drops has been proven as one of the effective strategies to address lactose intolerance. Commercial exogenous oral lactase is available in the market, which have been proven to alleviate the symptoms of Lactose intolerance. Experimental results have shown that co-ingestion of lactose and oral lactase is effective in reducing the breath hydrogen and clinical symptoms. Infantile colic is considered as a manifestation of early lactose intolerance and can be very exhausting for the child and parents. Oral lactase treatment in infantile colic have resulted in symptomatic relief and is now considered by medical practitioners (Narang & Shah, 2022). In rare cases, allergy to oral lactase has been reported (Binkley, 1996; Voisin & Borici-Mazi, 2016). Anaphylaxis due to oral lactase resulting into two cases (Lohrenz & Kanani, 2023; Voisin & Borici-Mazi, 2016). Some formats related to the oral lactase enzymes are given in Table 1.6.

1.5.4 Prebiotics and Probiotics

Probiotics are live microbial cells which when introduced, have a beneficial effect on the host. Probiotic cultures are tested and required to pass the gastrointestinal tract and reach and stabilize in the gut. Once established in the host intestine, they exhibit numerous health benefits. These functional benefits are mediated by various enzymes and molecules released by the probiotic culture *in-situ*. So, the ability of

Table 1.6 Oral lactase enzyme formats

S.No.	Enzyme delivery format	Findings	References
1	Chewable lactase tablet	Reduced the concentration of expired hydrogen and symptoms of LI	Sanders et al. (1992)
2	Lactase tablet	Co ingestion of lactose and lactase tablets significantly reduces both breath hydrogen and clinical symptoms associated with LI	Medow et al. (1990)
3	Chewable lactase tablet	Clinical symptoms like bloating, abdominal pain, diarrhea, flatulence, and borborygmi improved and hydrogen breath levels were reduced in lactose-intolerant patients	Baijal and Tandon (2021)
4	Acid resistant ß-galactosidase tablet	15,000 Units of ß-Galactosidase in lactose malabsorber individuals is effective in decreasing significantly the mean peak H_2 levels and the cumulative values of breath H_2 excretion in a small group of patients	Ibba et al. (2014))
5	Oral Lactase solution	Oral lactase treatment in infantile colic results in symptomatic relief	Narang and Shah (2022)
6	Pretreatment of feed with lactase	Transient lactose intolerance which is a causative factor in infant colic can be reduced by pre-treatment of feed with lactase	Kanabar et al. (2001)
7	Lactase tablet	Microbial endogenous lactase in yogurt was superior to exogenous commercial lactase tablet in alleviating lactose maldigestion.	Onwulata et al. (1989)
8	Lactase microcapsules	A powder of lactase microcapsules coated with HPMCP was found to be a suitable supplement for milk	Ahn et al. (2019))
9	Lactase drops	Lactase drops require prior incubation with milk formula to be effective	Kearney et al. (1998)
10	Lactogest (Softgel capsule), Lactaid (Caplet), Dairy Ease (Chewable)	All the three tablets, individually corresponding to approximately 6000 IU of lactase, were able to achieve effective hydrolysis of 20 g of lactose and less symptoms	Lin et al. (1993)

the probiotic cells, to control lactose intolerance, is stipulated from their ability to release lactase. The *in-situ* produced lactase metabolizes the dietary lactose, and prevents the secondary manifestations of free sugar in the intestine. Interventions related to prebiotics and probiotics and their associated actions are depicted in Table 1.7. Selected probiotic cultures have been proved to show positive effects on lactose intolerance induced symptoms (Li et al., 2023a, b).

1.6 Conclusion

Worldwide, the prevalence of lactose intolerance is very high. The digestive problems associated with the lactose intolerance are observed among many children and also in adults. However, the chronicity of the symptoms is depended on various

Table 1.7 Prebiotics and probiotics intervention with their associated actions

S. no.	Intervention	Associated actions	References
1	Probiotic yogurt with *Lactobacillus acidophilus* and *Bifidobacterium*	Probiotic yogurt fortified with *L. acidophilus* and *Bifidobacterium sp.* could safely and effectively decrease lactose intolerance symptoms and HBT	Masoumi et al. (2021)
2	Meta analysis of 12 studies on probiotic intake and effect on LI	Administration of probiotics was effective at improving adult lactose intolerance	Ahn et al. (2023)
3	Meta analysis of 5 studies on probiotic intake and effect on LI	*Limosilactobacillus reuteri* DSM 17938 and *Lactobacillus acidophilus* DDS-1 showed the best results in the management of LI symptoms. The prebiotic GOS (RP-G28) appeared to be more efficient in reducing post-treatment symptoms	de Oliveira et al. (2022)
4	Meta analysis of 5 studies on probiotic intake and effect on LI	Varying degrees of efficacy but an overall positive relationship between probiotics and LI in relation to specific strains and concentrations	Leis et al. (2020)
5	Probiotics for compensation of lactase deficiency	microbial β-galactosidase in the (fermented) milk product, delayed gastrointestinal transit, positive effects on intestinal functions and colonic microflora, and reduced sensitivity to symptoms	de Vrese et al. (2001)
6	Application of probiotic formulation (probiotic with β-galactosidase) to 8 female patients	Effectively reduced the frequency of symptoms (50%) of lactose intolerance including bloating, flatulence, etc. after 6 months study periods	Gingold-Belfer et al. (2020)
7	Randomized clinical control trials on 44 patients having lactose intolerance	*B. animalis* subsp. *animalis* IM386 and *L. plantarum* MP2026 containing products significantly reduced the symptoms of both groups including diarrhoea and flatulence	Roškar et al. (2017)
8	Efficacy of probiotic i3.1 in improving lactose intolerance symptoms compared with placebo after 8 weeks of treatment	i3.1 probiotic was safe and efficacious in reducing lactose intolerance symptoms in patients with lactose intolerance, but did not change the LHBT	Cano-Contreras et al. (2022)
9	Efficacy of a novel formulation of *Bifidobacterium longum* BB536 and *Lactobacillus rhamnosus* HN001 plus vitamin B6 (ZR) in 23 LI subjects with persistent symptoms during a lactose-free diet.	Selected probiotics and vitamin B6 to alleviate symptoms and gut dysbiosis in lactose intolerant patients with persistent functional gastrointestinal symptoms	Vitellio et al. (2019)

factors. However, the diagnosis of lactose intolerance is now a day quite simple, usually based on clinical grounds. Studies reported that lactose intolerant people can able to digests up to 5–10 g of lactose per day. However, it is essential to have better understanding of the intolerance caused by the lactose containing milk and milk products, so that we can consume milk in a better way, as complete negligence of milk products may cause to deficiency of several valuable nutrients of it. Hence, looking to the necessity for lactose intolerants, several lactose free products are being commercialized. Several microbial strains of lactase enzyme were discovered. Probiotics dairy products and plant-based drinks have been popularized as an alternative to avoid the lactose malabsorption. In conclusion, this chapter has delved into the complexities of lactose intolerance, and management strategies. By understanding the science behind lactose intolerance and adopting practical approaches, people can navigate their dietary habits and can improve their overall well-being. As we continue to unravel the mysteries of this condition, promoting awareness and education remains crucial for fostering a supportive environment for those affected by lactose malabsorption. Continued research will be instrumental in improving the strategies for diagnosis, treatment, and support for lactose intolerant dealing with this common digestive issue.

References

Abbasi, S., & Saeedabadian, A. (2015). Influences of lactose hydrolysis of milk and sugar reduction on some physical properties of ice cream. *Journal of Food Science and Technology, 52*(1), 367–374. https://doi.org/10.1007/s13197-013-1011-1

Ahn, S.-I., Lee, Y.-K., & Kwak, H.-S. (2019). Physicochemical and sensory properties of milk supplemented with lactase microcapsules coated with enteric coating materials. *Journal of Dairy Science, 102*(8), 6959–6970. https://doi.org/10.3168/jds.2018-15865

Ahn, S. I., Kim, M. S., Park, D. G., Han, B. K., & Kim, Y. J. (2023). Effects of probiotics administration on lactose intolerance in adulthood: A meta-analysis. *Journal of Dairy Science, 106*(7), 4489–4501. https://doi.org/10.3168/jds.2022-22762

Alharbi, O., & El-Sohemy, A. (2017). Lactose intolerance (LCT-13910C>T) genotype is associated with plasma 25-Hydroxyvitamin D concentrations in caucasians: A mendelian randomization study 1. *The Journal of Nutrition, 147*(6), 1063–1069. https://doi.org/10.3945/jn.116.246108

Amiri, M., Diekmann, L., von Köckritz-Blickwede, M., & Naim, H. Y. (2015). The diverse forms of lactose intolerance and the putative linkage to several cancers. *Nutrients, 7*(9), 7209–7230. https://doi.org/10.3390/nu7095332

Andrzej, P., Piotr, M., Borun, P., Skrzypczak-Zielinska, M., Wojciechowska-Lacka, A., Godlewski, D., & Banasiewicz, T. (2015). Influence of lactose intolerance on colorectal cancer incidence in the Polish population. *Hereditary Cancer in Clinical Practice, 13*(1), A7. https://doi.org/10.1186/1897-4287-13-S1-A7

Aragón, J. J., Hermida, C., Martínez-Costa, O. H., Sánchez, V., Martín, I., Sánchez, J. J., et al. (2014). Noninvasive diagnosis of hypolactasia with 4-Galactosylxylose (Gaxilose): A multicentre, open-label, phase IIB-III nonrandomized trial. *Journal of Clinical Gastroenterology, 48*(1), 29–36.

Arasi, S., Cafarotti, A., & Fiocchi, A. (2022). Cow's milk allergy. *Current Opinion in Allergy and Clinical Immunology, 22*(3), 181–187.

Araújo, N. G., Barbosa, I. M., Lima, T. L. S., Moreira, R. T., & Cardarelli, H. R. (2022). Development and characterization of lactose-free probiotic goat milk beverage with bioactive rich jambo pulp. *Journal of Food Science and Technology, 59*(10), 3806–3818. https://doi.org/10.1007/s13197-022-05399-z

Asrar, R., Farhan, M. H. R., Ahmad, M., Muni, R. F., Ilyas, I., Ihsan, H. M. T., Asrar, A., Shaukat, S., Aleem, M. T., Shaukat, I., & Shaukat, A. (2023). Lactose intolerance and emerging dairy allergies as public health perspective. In R. Z. Abbas, N. M. Saeed, M. Younus, L. Aguilar-Marcelino, & A. Khan (Eds.), *One health triad* (Vol. 2, pp. 138–146). Unique Scientific Publishers. https://doi.org/10.47278/book.oht/2023.53

Aysu, Ş., Akıl, F., Çevik, E., Kılıç, M. E., & İlyasoğlu, H. (2022). Development of homemade hazelnut milk-based beverage. *Journal of Culinary Science & Technology, 20*(5), 421–429. https://doi.org/10.1080/15428052.2020.1859033

Bácsi, K., Hitre, E., Kósa, J. P., Horváth, H., Lazáry, Á., Lakatos, P. L., Balla, B., Budai, B., Lakatos, P., & Speer, G. (2008). Effects of the lactase 13910 C/T and calcium-sensor receptor A986S G/T gene polymorphisms on the incidence and recurrence of colorectal cancer in Hungarian population. *BMC Cancer, 8*, 317. https://doi.org/10.1186/1471-2407-8-317

Baijal, R., & Tandon, R. K. (2021). Effect of lactase on symptoms and hydrogen breath levels in lactose intolerance: A crossover placebo-controlled study. *JGH Open, 5*(1), 143–148. https://doi.org/10.1002/jgh3.12463

Bianchini, C. B., Vieira, M. P. T., Arriola, N. D. A., Dias, C. O., Seraglio, S. K. T., Costa, A. C. O., Komatsu, R. A., Machado, B. D., Amboni, R. D. M. C., & Fritzen-Freire, C. B. (2020). Incorporation of uvaia (Eugenia pyriformis Cambess) pulp in yogurt: A promising application in the lactose-free dairy product market. *Journal of Food Processing and Preservation, 44*(10), e14829. https://doi.org/10.1111/jfpp.14829

Binkley, K. E. (1996). Allergy to supplemental lactase enzyme. *Journal of Allergy and Clinical Immunology, 97*(6), 1414–1416. https://doi.org/10.1016/S0091-6749(96)70213-0

Bøgh, K. L., & Larsen, J. M. (2021). Reducing allergenicity by proteolysis. In *Agents of change: Enzymes in milk and dairy products* (pp. 499–524). Springer.

Brew, K. (2003). Lactalbumin. In P. F. Fox & P. L. H. McSweeney (Eds.), *Advanced dairy chemistry, vol. 1, proteins* (3rd ed., pp. 387–419). Kluwer Academic/Plenum Publishers.

Cano-Contreras, A. D., Minero Alfaro, I. J., Medina López, V. M., Amieva Balmori, M., Remes Troche, J. M., Espadaler Mazo, J., & Perez Lopez, N. (2022). Efficacy of i3.1 probiotic on improvement of lactose intolerance symptoms: A randomized, placebo-controlled clinical trial. *Journal of Clinical Gastroenterology, 56*(2), 141–147. https://doi.org/10.1097/MCG.0000000000001456

Capcanari, T., Chirsanova, A., Covaliov, E., & Siminiuc, R. (2021). Development of lactose free yogurt technology for personalized nutrition. *Food and Nutrition Sciences, 12*(11), Article 11–Articl1135. https://doi.org/10.4236/fns.2021.1211082

Caroli, A., Poli, A., Ricotta, D., Banfi, G., & Cocchi, D. (2011). Invited review: Dairy intake and bone health: A viewpoint from the state of the art1. *Journal of Dairy Science, 94*(11), 5249–5262. https://doi.org/10.3168/jds.2011-4578

Casellas, F., Aparici, A., Pérez, M. J., & Rodríguez, P. (2016). Perception of lactose intolerance impairs health-related quality of life. *European Journal of Clinical Nutrition, 70*(9), Article 9–Artic1072. https://doi.org/10.1038/ejcn.2016.80

Catanzaro, R., Sciuto, M., Singh, B., Pathak, S., & Marotta, F. (2020). Irritable bowel syndrome and lactose intolerance: The importance of differential diagnosis. A monocentric study. *Minerva Gastroenterology, 67*(1), 72–78. https://doi.org/10.23736/S1121-421X.20.02734-8

Chan, G. M., Hoffman, K., & McMurry, M. (1995). Effects of dairy products on bone and body composition in pubertal girls. *The Journal of Pediatrics, 126*(4), 551–556. https://doi.org/10.1016/S0022-3476(95)70348-9

Cincotta, F., Condurso, C., Tripodi, G., Merlino, M., Prestia, O., Stanton, C., & Verzera, A. (2021). Comparison of lactose free and traditional mozzarella cheese during shelf-life by aroma compounds and sensory analysis. *LWT, 140*, 110845. https://doi.org/10.1016/j.lwt.2020.110845

Connor, F., Silvia, S., Enza, D. A., Baldassarre, M. E., Miriam, A., Di Bella, G., et al. (2022). Cows' milk allergy-associated constipation: When to look for it?. A narrative review. *Nutrients, 14*, 1317.
Craveiro Barra, S. N., Gomes, P., & Leitão Marques, A. (2012). Severe lactose intolerance in a patient with coronary artery disease and ischemic cardiomyopathy. *Revista Portuguesa de Cardiologia, 31*(12), 821–824. https://doi.org/10.1016/j.repc.2012.04.014
Cui, L., Jia, Q., Zhao, J., Hou, D., & Zhou, S. (2023). A comprehensive review on oat milk: From oat nutrients and phytochemicals to its processing technologies, product features, and potential applications. *Food & Function, 14*(13), 5858–5869. https://doi.org/10.1039/D3FO00893B
Cutrim, C. S., de Barros, R. F., da Costa, M. P., Franco, R. M., Conte-Junior, C. A., & Cortez, M. A. S. (2016). Survival of Escherichia coli O157:H7 during manufacture and storage of traditional and low lactose yogurt. *LWT, 70*, 178–184. https://doi.org/10.1016/j.lwt.2016.02.047
Czyzewska, K., & Trusek, A. (2021). Encapsulated NOLA™ Fit 5500 Lactase–An economically beneficial way to obtain Lactose-Free milk at low temperature. *Catalysts, 11*(5), Article 5. https://doi.org/10.3390/catal11050527
da Silva, F. I., Souza, F. A., Ruschel, J., Badaró, A. C. L., Tonial, I. B., de Castro-Cislaghi, F. P., Quast, L. B., Bertan, L. C., & Quast, E. (2020). Production of naturally "lactose free" fresh cheese. *Research, Society and Development, 9*(10), Article 10. https://doi.org/10.33448/rsd-v9i10.8590
de Medeiros, A. C., Tavares Filho, E. R., & Bolini, H. M. A. (2021). Temporal profile of low calorie lactose-free ice cream chocolate flavor: Temporal dominance sensation and multiple time–intensity analysis. *Journal of Food Science and Technology, 58*(8), 3164–3173. https://doi.org/10.1007/s13197-020-04819-2
de Oliveira, L. S., Wendt, G. W., Crestani, A. P. J., & Casaril, K. B. P. B. (2022). The use of probiotics and prebiotics can enable the ingestion of dairy products by lactose intolerant individuals. *Clinical Nutrition, 41*(12), 2644–2650. https://doi.org/10.1016/j.clnu.2022.10.003
de Vrese, M., Stegelmann, A., Richter, B., Fenselau, S., Laue, C., & Schrezenmeir, J. (2001). Probiotics–Compensation for lactase insufficiency. *The American Journal of Clinical Nutrition, 73*(2 Suppl), 421S–429S. https://doi.org/10.1093/ajcn/73.2.421s
Dekker, P. J. T., Koenders, D., & Bruins, M. J. (2019). Lactose-free dairy products: Market developments, production, nutrition and health benefits. *Nutrients, 11*(3), Article 3. https://doi.org/10.3390/nu11030551
Deng, Y., Misselwitz, B., Dai, N., & Fox, M. (2015). Lactose intolerance in adults: biological mechanism and dietary management. *Nutrients, 7*(9), 8020–8035.
Di Stefano, M., Veneto, G., Malservisi, S., Cecchetti, L., Minguzzi, L., Strocchi, A., & Corazza, G. R. (2002). Lactose malabsorption and intolerance and peak bone mass. *Gastroenterology, 122*(7), 1793–1799. https://doi.org/10.1053/gast.2002.33600
Domínguez Jiménez, J. L., & Fernández Suárez, A. (2016). Correlation between capillary and venous blood glucose in the lactose tolerance test. *Digestive Diseases and Sciences, 61*, 208–214.
Doreau, M., & Boulot, S. (1989). Recent knowledge on mare milk production: A review. *Livestock Production Science, 22*(3–4), 213–235.
Enattah, N. S., Sahi, T., Savilahti, E., Terwilliger, J. D., Peltonen, L., & Järvelä, I. (2002). Identification of a variant associated with adult-type hypolactasia. *Nature Genetics, 30*(2), 233–237.
Facioni, M. S., Dominici, S., Marescotti, F., Covucci, R., Taglieri, I., Venturi, F., & Zinnai, A. (2021). Lactose residual content in PDO cheeses: Novel inclusions for consumers with lactose intolerance. *Food, 10*(9), Article 9. https://doi.org/10.3390/foods10092236
Fox, P. F., Uniacke-Lowe, T., McSweeney, P. L., & O'Mahony, J. A. (2015). Milk proteins. In *Dairy chemistry and biochemistry* (pp. 145–238). Springer.
Franzè, A., & Bertelè, A. (2010). Intolleranza al lattosio nella pratica clinica. *Rivista della Società Italiana di Medicina Generale, 3*, 36–40.

Fuhrmann, G., & Leroux, J.-C. (2014). Improving the stability and activity of oral therapeutic enzymes–Recent advances and perspectives. *Pharmaceutical Research, 31*(5), 1099–1105. https://doi.org/10.1007/s11095-013-1233-y

Gänzle, M. G., Haase, G., & Jelen, P. (2008). Lactose: Crystallization, hydrolysis and value-added derivatives. *International Dairy Journal, 18*(7), 685–694.

Gingold-Belfer, R., Levy, S., Layfer, O., Pakanaev, L., Niv, Y., Dickman, R., & Perets, T. T. (2020). Use of a novel probiotic formulation to alleviate lactose intolerance symptoms-a pilot study. *Probiotics and Antimicrobial Proteins, 12*(1), 112–118. https://doi.org/10.1007/s12602-018-9507-7

Harju, M., Kallioinen, H., & Tossavainen, O. (2012). Lactose hydrolysis and other conversions in dairy products: Technological aspects. *International Dairy Journal, 22*(2), 104–109. https://doi.org/10.1016/j.idairyj.2011.09.011

Harrington, L. K., & Mayberry, J. F. (2008). A re-appraisal of lactose intolerance. *International Journal of Clinical Practice, 62*(10), 1541–1546.

Houben, E., De Preter, V., Billen, J., Van Ranst, M., & Verbeke, K. (2015). Additional value of CH4 measurement in a combined 13C/H2 lactose malabsorption breath test: A retrospective analysis. *Nutrients, 7*(9), 7469–7485.

Ibba, I., Gilli, A., Boi, M. F., & Usai, P. (2014). Effects of exogenous lactase administration on hydrogen breath excretion and intestinal symptoms in patients presenting lactose malabsorption and intolerance. *BioMed Research International, 2014*, e680196. https://doi.org/10.1155/2014/680196

Ilich, J. Z., & Kerstetter, J. E. (2000). Nutrition in bone health revisited: A story beyond calcium. *Journal of the American College of Nutrition, 19*(6), 715–737. https://doi.org/10.1080/07315724.2000.10718070

Infante, D., & Tormo, R. (2000). Risk of inadequate bone mineralization in diseases involving long-term suppression of dairy products. *Journal of Pediatric Gastroenterology and Nutrition, 30*(3), 310–313.

Kanabar, D., Randhawa, M., & Clayton, P. (2001). Improvement of symptoms in infant colic following reduction of lactose load with lactase. *Journal of Human Nutrition and Dietetics: The Official Journal of the British Dietetic Association, 14*(5), 359–363. https://doi.org/10.1046/j.1365-277x.2001.00304.x

Kaur, K., Mahmood, S., & Mahmood, A. (2006). Hypolactasia as a molecular basis of lactose intolerance. *Indian Journal of Biochemistry and Biophysics, 43*(5), 267–274.

Kearney, M., Hayes, C., & Hyland. (1998). A trial of lactase in the management of infant colic. *Journal of Human Nutrition and Dietetics, 11*(4), 281–285. https://doi.org/10.1046/j.1365-277X.1998.00108.x

Kim, J.-W. (2017). Lactose intolerance and colorectal cancer. *Annals of Coloproctology, 33*(5), 157–158. https://doi.org/10.3393/ac.2017.33.5.157

Kocabaş, H., Ergin, F., Aktar, T., & Küçükçetin, A. (2022). Effect of lactose hydrolysis and salt content on the physicochemical, microbiological, and sensory properties of ayran. *International Dairy Journal, 129*, 105360. https://doi.org/10.1016/j.idairyj.2022.105360

Kudlacek, S., Freudenthaler, O., Weissböeck, H., Schneider, B., & Willvonseder, R. (2002). Lactose intolerance: A risk factor for reduced bone mineral density and vertebral fractures? *Journal of Gastroenterology, 37*(12), 1014–1019. https://doi.org/10.1007/s005350200171

Kun, Z., Greenfield, H., Xueqin, D., & Fraser, D. R. (2001). Improvement of bone health in childhood and adolescence. *Nutrition Research Reviews, 14*(1), 119–152. https://doi.org/10.1079/NRR200120

Kundu, P., Dhankhar, J., & Sharma, A. (2018). Development of non dairy milk alternative using soymilk and almond milk. *Current Research in Nutrition and Food Science Journal, 6*(1), 203–210.

Lambrini, K., Aikaterini, F., Konstantinos, K., Christos, I., Ioanna, P. V., & Areti, T. (2021). Milk nutritional composition and its role in human health. *Journal of Pharmacy and Pharmacology, 9*, 8–13.

Leis, R., de Castro, M.-J., de Lamas, C., Picáns, R., & Couce, M. L. (2020). Effects of prebiotic and probiotic supplementation on lactase deficiency and lactose intolerance: A systematic review of controlled trials. *Nutrients, 12*(5), Article 5. https://doi.org/10.3390/nu12051487

Li, A., Zheng, J., Han, X., Jiang, Z., Yang, B., Yang, S., Zhou, W., Li, C., & Sun, M. (2023a). Health implication of lactose intolerance and updates on its dietary management. *International Dairy Journal, 140*, 105608.

Li, A., Zheng, J., Han, X., Yang, S., Cheng, S., Zhao, J., Zhou, W., & Lu, Y. (2023b). Advances in low-Lactose/Lactose-Free dairy products and their production. *Food, 12*(13), Article 13. https://doi.org/10.3390/foods12132553

Lin, M. Y., Dipalma, J. A., Martini, M. C., Gross, C. J., Harlander, S. K., & Savaiano, D. A. (1993). Comparative effects of exogenous lactase (β-galactosidase) preparations onin vivo lactose digestion. *Digestive Diseases and Sciences, 38*(11), 2022–2027. https://doi.org/10.1007/BF01297079

Lohrenz, S. K., & Kanani, A. (2023). The cow's milk allergy that wasn't: Allergy to supplemental oral lactase enzyme. *Allergy, Asthma & Clinical Immunology, 19*(1), 61. https://doi.org/10.1186/s13223-023-00809-z

Lomer, M. C., Parkes, G. C., & Sanderson, J. D. (2008). lactose intolerance in clinical practice–myths and realities. *Alimentary Pharmacology & Therapeutics, 27*(2), 93–103. https://doi.org/10.1111/j.1365-2036.2007.03557.x

Maleki, N., Khodaiyan, F., & Mousavi, S. M. (2015). Antioxidant activity of fermented Hazelnut milk. *Food Science and Biotechnology, 24*(1), 107–115. https://doi.org/10.1007/s10068-015-0016-0

Martínez Rodríguez, M., de Samaniego-Vaesken, M. L., & Alonso-Aperte, E. (2021). A new food composition database of lactose-free products commercialized in Spain: Differences in nutritional composition as compared to traditional products. *Food, 10*(4), Article 4. https://doi.org/10.3390/foods10040851

Masoumi, S. J., Mehrabani, D., Saberifiroozi, M., Fattahi, M. R., Moradi, F., & Najafi, M. (2021). The effect of yogurt fortified with Lactobacillus acidophilus and Bifidobacterium sp. Probiotic in patients with lactose intolerance. *Food Science & Nutrition, 9*(3), 1704–1711. https://doi.org/10.1002/fsn3.2145

Mattar, R., de Campos Mazo, D. F., & Carrilho, F. J. (2012). Lactose intolerance: diagnosis, genetic, and clinical factors. *Clinical and Experimental Gastroenterology, 5*, 113–121. https://doi.org/10.2147/ceg.s32368

Matthews, S. B., Waud, J. P., Roberts, A. G., & Campbell, A. K. (2005). Systemic lactose intolerance: a new perspective on an old problem. *Postgraduate Medical Journal, 81*(953), 167–173.

Medow, M. S., Thek, K. D., Newman, L. J., Berezin, S., Glassman, M. S., & Schwarz, S. M. (1990). β-Galactosidase Tablets in the Treatment of Lactose Intolerance in Pediatrics. *American Journal of Diseases of Children, 144*(11), 1261–1264. https://doi.org/10.1001/archpedi.1990.02150350093034

Misselwitz, B., Butter, M., Verbeke, K., & Fox, M. R. (2019). Update on lactose malabsorption and intolerance: pathogenesis, diagnosis and clinical management. *Gut, 68*, 2080–2091.

Monti, L., Negri, S., Meucci, A., Stroppa, A., Galli, A., & Contarini, G. (2017). Lactose, galactose and glucose determination in naturally "lactose free" hard cheese: HPAEC-PAD method validation. *Food Chemistry, 220*, 18–24. https://doi.org/10.1016/j.foodchem.2016.09.185

Morelos-Gomez, A., Terashima, S., Yamanaka, A., Cruz-Silva, R., Ortiz-Medina, J., Sánchez-Salas, R., Fajardo-Díaz, J. L., Muñoz-Sandoval, E., López-Urías, F., Takeuchi, K., Tejima, S., Terrones, M., & Endo, M. (2021). Graphene oxide membranes for lactose-free milk. *Carbon, 181*, 118–129. https://doi.org/10.1016/j.carbon.2021.05.005

Murshed, M. (2018). Mechanism of bone mineralization. *Cold Spring Harbor Perspectives in Medicine, 8*(12), a031229. https://doi.org/10.1101/cshperspect.a031229

Nakayama, T., & Amachi, T. (2002). β-Galactosidase, enzymology. In *Encyclopedia of bioprocess technology*. Wiley. https://doi.org/10.1002/0471250589.ebt102

Narang, M., & Shah, D. (2022). Oral lactase for infantile colic: A randomized double-blind placebo-controlled trial. *BMC Pediatrics, 22*(1), 468. https://doi.org/10.1186/s12887-022-03531-8

Obermayer-Pietsch, B. M., Bonelli, C. M., Walter, D. E., Kuhn, R. J., Fahrleitner-Pammer, A., Berghold, A., Goessler, W., Stepan, V., Dobnig, H., Leb, G., & Renner, W. (2004). Genetic predisposition for adult lactose intolerance and relation to diet, bone density, and bone fractures. *Journal of Bone and Mineral Research, 19*(1), 42–47. https://doi.org/10.1359/jbmr.0301207

Oftedal, O. T., & Jenness, R. (1988). Interspecies variation in milk composition among horses, zebras and asses (Perissodactyla: Equidae). *Journal of Dairy Research, 55*(1), 57–66.

Onwulata, C. I., Rao, D. R., & Vankineni, P. (1989). Relative efficiency of yogurt, sweet acidophilus milk, hydrolyzed-lactose milk, and a commercial lactase tablet in alleviating lactose maldigestion. *The American Journal of Clinical Nutrition, 49*(6), 1233–1237. https://doi.org/10.1093/ajcn/49.6.1233

Pachekrepapol, U., Somboonchai, N., & Krimjai, W. (2021). Physicochemical, rheological, and microbiological properties of lactose-free functional yogurt supplemented with fructooligosaccharides. *Journal of Food Processing and Preservation, 45*(1), e15017. https://doi.org/10.1111/jfpp.15017

Palacios, C. (2006). The role of nutrients in bone health, from A to Z. *Critical Reviews in Food Science and Nutrition, 46*(8), 621–628. https://doi.org/10.1080/10408390500466174

Pansuwan, Y. (2023). Lactose intolerance: Biochemistry perspective. *Greater Mekong Subregion Medical Journal, 3*(2), 101–108.

Parmar, S. S., Singh, S., & Sharma, R. S. (1989). Compositional and quality characteristics of paneer made from soya beans and buffalo milk under different heat treatments. *Journal of the Science of Food and Agriculture, 47*(4), 463–473. https://doi.org/10.1002/jsfa.2740470407

Pophaly, S. D., Tiwari, S., Tripathi, A. K., & Manorama. (2022). Effect of milk and dairy products in colorectal cancer. In D. Shukla, N. K. Vishvakarma, & G. P. Nagaraju (Eds.), *colon cancer diagnosis and therapy* (Vol. 3, pp. 325–337). Springer International Publishing. https://doi.org/10.1007/978-3-030-72702-4_15

Popovici, C., Baerle, A., & Tatarov, P. (2017). *Biochemical aspects of walnut dairy free milk*. http://repository.utm.md/handle/5014/18114.

Rasinperä, H., Forsblom, C., Enattah, N. S., Halonen, P., Salo, K., Victorzon, M., Mecklin, J.-P., Järvinen, H., Enholm, S., Sellick, G., Alazzouzi, H., Houlston, R., Robinson, J., Groop, P.-H., Tomlinson, I., Schwartz, S., Aaltonen, L. A., & Järvelä, I. (2005). The C/C−13910 genotype of adult-type hypolactasia is associated with an increased risk of colorectal cancer in the Finnish population. *Gut, 54*(5), 643–647. https://doi.org/10.1136/gut.2004.055939

Rezaie, A., Buresi, M., Lembo, A., Lin, H., McCallum, R., Rao, S., & Pimentel, M. (2017). Hydrogen and methane-based breath testing in gastrointestinal disorders: The North American consensus. *The American Journal of Gastroenterology, 112*(5), 775–784. https://doi.org/10.1038/ajg.2017.46

Rizzoli, R. (2022). Dairy products and bone health. *Aging Clinical and Experimental Research, 34*(1), 9–24. https://doi.org/10.1007/s40520-021-01970-4

Robles, L., & Priefer, R. (2020). Lactose intolerance: What your breath can tell you. *Diagnostics, 10*(6), 412. https://doi.org/10.3390/diagnostics10060412

Roškar, I., Švigelj, K., Štempelj, M., Volfand, J., Štabuc, B., Malovrh, Š., & Rogelj, I. (2017). Effects of a probiotic product containing Bifidobacterium animalis subsp. animalis IM386 and Lactobacillus plantarum MP2026 in lactose intolerant individuals: Randomized, placebo-controlled clinical trial. *Journal of Functional Foods, 35*, 1–8. https://doi.org/10.1016/j.jff.2017.05.020

Roth, G. A., Mensah, G. A., Johnson, C. O., Addolorato, G., Ammirati, E., Baddour, L. M., Barengo, N. C., Beaton, A. Z., Benjamin, E. J., Benziger, C. P., Bonny, A., Brauer, M., Brodmann, M., Cahill, T. J., Carapetis, J., Catapano, A. L., Chugh, S. S., Cooper, L. T., Coresh, J., et al. (2020). Global burden of cardiovascular diseases and risk factors, 1990–2019. *Journal of the American College of Cardiology, 76*(25), 2982–3021. https://doi.org/10.1016/j.jacc.2020.11.010

Sabra, A., Bellanti, J. A., Rais, J. M., Castro, H. J., de Inocencio, J. M., & Sabra, S. (2003). IgE and non-IgE food allergy. *Annals of Allergy, Asthma & Immunology, 90*(6), 71–76.

Sanders, S. W., Tolman, K. G., & Reitberg, D. P. (1992). Effect of a single dose of lactase on symptoms and expired hydrogen after lactose challenge in lactose-intolerant subjects. *Clinical Pharmacy, 11*(6), 533–538.

Schnedl, W. J., Meier-Allard, N., Lackner, S., Enko, D., Mangge, H., & Holasek, S. J. (2020). Increasing expiratory hydrogen in lactose intolerance is associated with additional food intolerance/malabsorption. *Nutrients, 12*(12), 3690. https://doi.org/10.3390/nu12123690

Segall, J. J. (1994). Dietary lactose as a possible risk factor for ischaemic heart disease: Review of epidemiology. *International Journal of Cardiology, 46*(3), 197–207. https://doi.org/10.1016/0167-5273(94)90242-9

Segall, J. J. (2008). Hypothesis: Is lactose a dietary risk factor for ischaemic heart disease? *International Journal of Epidemiology, 37*(6), 1204–1208. https://doi.org/10.1093/ije/dyn169

Shafi, A., & Husain, Q. (2022). Intolerance to milk lactose, diagnostic tests and dietary management: A recent update. *Avicenna Journal of Medical Biochemistry, 10*(1), 71–81. https://doi.org/10.34172/ajmb.2022.10

Sicherer, S. H., & Sampson, H. A. (2018). Food allergy: A review and update on epidemiology, pathogenesis, diagnosis, prevention, and management. *Journal of Allergy and Clinical Immunology, 141*(1), 41–58.

Singh, P., Rao, P. S., Sharma, V., & Arora, S. (2021). Physico-chemical aspects of lactose hydrolysed milk system along with detection and mitigation of maillard reaction products. *Trends in Food Science & Technology, 107*, 57–67. https://doi.org/10.1016/j.tifs.2020.11.030

Skryplonek, K., Gomes, D., Viegas, J., Pereira, C., & Henriques, M. (2017). Lactose-free frozen yogurt: Production and characteristics. *Acta Scientiarum Polonorum Technologia Alimentaria, 16*(2), 171–179. https://doi.org/10.17306/J.AFS.2017.0478

Skryplonek, K., Henriques, M., Gomes, D., Viegas, J., Fonseca, C., Pereira, C., Dmytrów, I., & Mituniewicz-Małek, A. (2019). Characteristics of lactose-free frozen yogurt with κ-carrageenan and corn starch as stabilizers. *Journal of Dairy Science, 102*(9), 7838–7848. https://doi.org/10.3168/jds.2019-16556

Solomons, N. W. (2002). Fermentation, fermented foods and lactose intolerance. *European Journal of Clinical Nutrition, 56*(4), Article 4. https://doi.org/10.1038/sj.ejcn.1601663

Suchy, F. J., Brannon, P. M., Carpenter, T. O., Fernandez, J. R., Gilsanz, V., Gould, J. B., Hall, K., Hui, S. L., Lupton, J., Mennella, J., Miller, N. J., Osganian, S. K., Sellmeyer, D. E., & Wolf, M. A. (2010). NIH consensus development conference statement: Lactose intolerance and health. *NIH Consensus and State-of-the-Science Statements, 27*(2), 1–27.

Szilagyi, A., & Ishayek, N. (2018). Lactose intolerance, dairy avoidance, and treatment options. *Nutrients, 10*(12), 1994.

Tarabra, E., Pazienza, P., Borghesio, E., Actis, G. C., Tappero, G., Framarin, L., Ayoubi, M., Castellino, F., Leone, N., Sansoè, G., De Paolis, P., Comandone, A., & Rosina, F. (2010). LCT-13910C>T polymorphism-associated lactose malabsorption and risk for colorectal cancer in Italy. *Digestive and Liver Disease, 42*(10), 741–743. https://doi.org/10.1016/j.dld.2010.02.013

Ugidos-Rodríguez, S., Matallana-González, M. C., & Sánchez-Mata, M. C. (2018). Lactose malabsorption and intolerance: A review. *Food & Function, 9*(8), 4056–4068. https://doi.org/10.1039/C8FO00555A

van den Heuvel, E. G. H. M., & Steijns, J. M. J. M. (2018). Dairy products and bone health: How strong is the scientific evidence? *Nutrition Research Reviews, 31*(2), 164–178. https://doi.org/10.1017/S095442241800001X

Vandenplas, Y. (2015). Lactose intolerance. *Asia Pacific Journal of Clinical Nutrition, 24*(Supplement), S9–S13. https://doi.org/10.6133/apjcn.2015.24.s1.02

Vanga, S. K., & Raghavan, V. (2018). How well do plant based alternatives fare nutritionally compared to cow's milk? *Journal of Food Science and Technology, 55*(1), 10–20. https://doi.org/10.1007/s13197-017-2915-y

Vargas, A. J., & Thompson, P. A. (2012). Diet and nutrient factors in colorectal cancer risk. *Nutrition in Clinical Practice: Official Publication of the American Society for Parenteral and Enteral Nutrition, 27*(5), 613–623. https://doi.org/10.1177/0884533612454885

Varju, P., Ystad, B., Gede, N., Hegyi, P., Pecsi, D., & Czimmer, J. (2020). The role of small intestinal bacterial overgrowth and false positive diagnosis of lactose intolerance in southwest Hungary–A retrospective observational study. *PLoS One, 15*(5), e0230784. https://doi.org/10.1371/journal.pone.0230784

Vázquez, S. M., de Rojas, J. N., Troche, J. R., Adame, E. C., Ruíz, R. R., & Domínguez, L. U. (2020). The importance of lactose intolerance in individuals with gastrointestinal symptoms. *Revista de Gastroenterología de México (English Edition), 85*(3), 321–331.

Vinther, S., Rumessen, J. J., & Christensen, M. (2015). Pharmaceutical drugs containing lactose can as a rule be used by persons with lactose intolerance. *Ugeskrift for Laeger, 177*(11), V11140608.

Vitellio, P., Celano, G., Bonfrate, L., Gobbetti, M., Portincasa, P., & De Angelis, M. (2019). Effects of bifidobacterium longum and lactobacillus rhamnosus on gut microbiota in patients with lactose intolerance and persisting functional gastrointestinal symptoms: A randomised, double-blind, cross-over study. *Nutrients, 11*(4), Article 4. https://doi.org/10.3390/nu11040886

Voisin, M. R., & Borici-Mazi, R. (2016). Anaphylaxis to supplemental oral lactase enzyme. *Allergy, Asthma & Clinical Immunology, 12*(1), 66. https://doi.org/10.1186/s13223-016-0171-8

Walsh, J., Meyer, R., Shah, N., Quekett, J., & Fox, A. T. (2016). Differentiating milk allergy (IgE and non-IgE mediated) from lactose intolerance: understanding the underlying mechanisms and presentations. *British Journal of General Practice, 66*(649), 609–611.

Yang, W., Liu, L., Masugi, Y., Qian, Z. R., Nishihara, R., Keum, N., Wu, K., Smith-Warner, S., Ma, Y., Nowak, J. A., Momen-Heravi, F., Zhang, L., Bowden, M., Morikawa, T., da Silva, A., Wang, M., Chan, A. T., Fuchs, C. S., Meyerhardt, J. A., et al. (2018). Calcium intake and risk of colorectal cancer according to expression status of calcium-sensing receptor (CASR). *Gut, 67*(8), 1475–1483. https://doi.org/10.1136/gutjnl-2017-314163

Zadow, J. G. (1986). *Lactose hydrolysed dairy products*. https://publications.csiro.au/rpr/pub?list=BRO&pid=procite:40653ac1-0c6c-40d9-9afb-c106d8759458.

Zhang, H., Tao, Y., He, Y., Pan, J., Yang, K., Shen, J., & Gao, C. (2020). Preparation of Low-lactose milk powder by coupling membrane technology. *ACS Omega, 5*(15), 8543–8550. https://doi.org/10.1021/acsomega.9b04252

Zhu, Y., Wang, P. P., Zhai, G., Bapat, B., Savas, S., Woodrow, J. R., Sharma, I., Li, Y., Zhou, X., Yang, N., Campbell, P. T., Dicks, E., Parfrey, P. S., & Mclaughlin, J. R. (2017). Vitamin D receptor and calcium-sensing receptor polymorphisms and colorectal cancer survival in the Newfoundland population. *British Journal of Cancer, 117*(6), Article 6. https://doi.org/10.1038/bjc.2017.242

Chapter 2
Market for Lactose Hydrolyzed Dairy Products in India: Status and Prospects

G. P. Shivaswamy, Laxmana Naik N., and M. Sivaram

2.1 Introduction

Milk is a fascinating fluid; milk and dairy products are the important constituents of the human daily diet since the times of domestication (Silanikove et al., 2015). Milk provides an important array of nutrient ingredients including high quality proteins, calcium, lactose, fat and many other essential nutrients in almost ideal proportion which are necessary for health. Hundreds of ingredients can be prepared from the single source 'milk' and these ingredients can be used in thousands of food and pharmaceutical applications. Lactose is the principal carbohydrate of mammal milk, with few exceptions (e.g., sea lions and walruses), and it is the source of nourishment for newborns (Misselwitz et al., 2019). Lactose usually occurs in milk as free form but a small amount occurs in the form of lactose-containing oligosaccharides (Morrissey, 1985). After weaning, mammals are not capable of digesting milk, and humans are the only species that continue to consume and digest milk (Ségurel & Bon, 2017) because of a phenomenon of gene–culture co-evolution. An enzyme lactase (also known as β-Galactosidase or lactase-phlorizin hydrolase) found in the lining of the small intestine, it specifically hydrolyzes the lactose into two simpler forms of sugar such as glucose and galactose (Malik & Panuganti, 2023; Hernandez et al., 2024). The body then absorbs these simpler sugars into the bloodstream. However, lactase deficiency and lactose malabsorption may lead to lactose intolerance. Many people can manage the symptoms of lactose intolerance by consumption of low-lactose milk or lactose free milk or a wide range of lactose-free dairy

G. P. Shivaswamy (✉) · M. Sivaram
Agricultural Economics, Southern Regional Station, ICAR-National Dairy Research Institute, Bengaluru, Karnataka, India

L. Naik N.
Dairy Chemistry Section, Southern Regional Station, ICAR-National Dairy Research Institute, Bengaluru, Karnataka, India

P. S. Rao et al. (eds.), *Lactose Hydrolysis in Dairy Products*,
https://doi.org/10.1007/978-3-031-78207-7_2

products; in which the lactose is hydrolyzed depending on the processing conditions (Suchy et al., 2010; Vardhanabhuti & Wang, 2022). Lactose hydrolyzed dairy products are preferred by health-conscious consumers because of numerous health benefits of milk nutrients and reduced caloric profile. In this context, this chapter highlights status and drivers of market potential for lactose hydrolyzed dairy products.

2.2 Status of Lactose Hydrolyzed Dairy Products

2.2.1 Global Status

Lactose free dairy products is witnessing a fastest growth among global dairy products. Figure 2.1 shows the global market share of different lactose free products. Consumers' demand for health-enhancing food products has grown rapidly over the past decades (Santeramo et al., 2018). A consumer study conducted in Germany reports that the influence of consumers' involvement in purchasing decisions of dietary supplements compared to ordinary food products (Kamrath et al., 2019). The Food and Drug Administration sets specific rules for what food manufacturers can call "light," "low," "reduced," "free," and other terms. A nutrient comparative claims related to these terms may try to convince the consumer on the food choice (Randell, 2010; Laxmana et al., 2018). One of the earliest studies reported three separate theoretical prospective predictions to assess the consumer purchasing psychological behaviour. The attribution theory relates to the product odour and taste and consumer be more likely to purchase the products (Lammers, 1991). In the

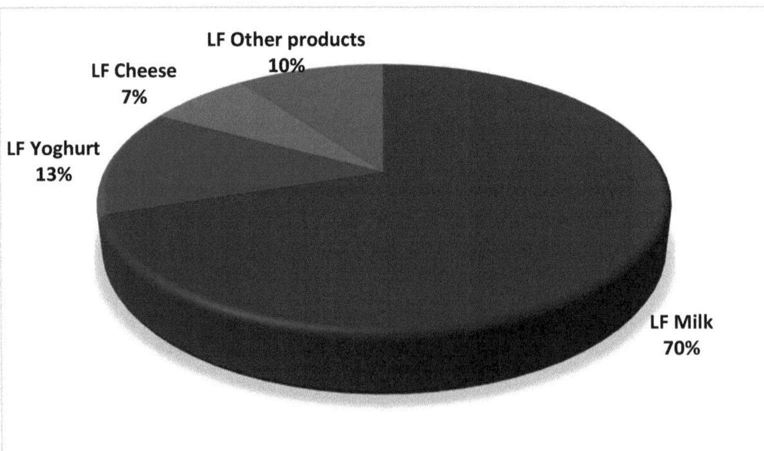

Fig. 2.1 Global market share of major lactose free products in 2017. (Source: Authors calculation using data from Euromonitor Database, 2019)

same prospective, consumer opinion about 'free-form' is rapidly evolving, as a result, the sales of lactose hydrolyzed dairy food products and their ingredients are growing exponentially.

The global lactose hydrolyzed dairy products market segment has shown a significant growth in recent years. Liquid milk is the largest category of lactose-free dairy, represents two-thirds of the market turnover and drives the absolute growth of the category. The second largest category is lactose-free yogurt, has a market share of around 13–15%. Lactose-free cheese accounts for 7% market turnover. Western Europe is the biggest and fastest growing lactose-free market, followed by Latin America (Dekker et al., 2019). Regional analysis on global lactose free dairy products market indicated that, geographically, North America (40%) and Europe (20%) have emerged as prominent players in lactose-free market segment, while the Asia-Pacific is gaining momentum at 15%. South America contributes to10% and Middle East Africa around 15%. The top market players operating in the global lactose-free dairy industry include Arla Foods, Danone S.A., Gujrat Cooperative Milk Marketing Federation (GCMMF), Johnson & Johnson Services, Inc., Lifeway Foods, Inc., Nestle S.A., Organic Valley, Saputo, Inc., and Smithfoods, Inc. Prairie Farms Dairy, Fonterra Group, Lala U.S., Lactaid Inc. Monmilk from China, Fairlife, and others (Allied Market research, 2022).

The global market of lactose free dairy products was valued at over USD 14.2 billion in 2022. It is expected to grow at 6.17% during 2023–2030 to reach USD 21.59 billion in 2030 (Market Research Future, 2024). Allied Market Research data indicates that the global lactose-free dairy market is projected to reach $24.36 billion by 2031 with a compound annual growth rate (CAGR) of 8% from 2022 to 2031 (Allied Market research, 2022). Global lactose free dairy products market is anticipated to exhibit a CAGR of 7.12% from 2023 to 2033, with a projected value of $29,158.8 million by 2033 (BIS Research, 2024). The global lactose and derivatives market is expected to reach the value of USD 72.95 billion by 2029, at a CAGR of 7.80% during the forecast period of 2022 to 2029 (Data bridge market research, 2022). Market for commercially available lactase enzymes is also growing drastically. The global Lactase enzyme market is expected to witness a CAGR of 6.3% from 2022 to 2027 (Markets and Markets, 2022). Prominent market players of Lactase enzymes with their brands /trade names are listed below: Astrolact by Calza Clemente, Italy; Biolactase by Kerry Bioscience, Ireland; Enzeco Fungal Lactase by Enzyme Development Corporation, New York, USA; GODO-YNL2 Lactase by DuPont Danisco, Copenhagen, Denmark; Lactase by Anthem Biosciences, India; Lactase by Bio-Cat Inc., USA; Lactase by Sisco Research Laboratories, India; Lactomax by Prozyn, Sao Paulo, Brazil;; Lactozym® by Novozymes, Denmark; Maxilact® by DSM Food Specialties, Netherlands; NOLA™ Fit and Ha-Lactase™ by CHR Hansen, Denmark; Validase fungal lactase concentrate and Validase yeast lactase by Valley Research, Inc., South Bend, USA.

2.2.2 Scenario in India

India is indisputably the world's biggest dairy industry in terms of milk production. During 2022–23, India produced close to 230.58 million tonnes of milk, accounting for 25% of total world's milk production (GOI, 2024). In India, dairy products are the integral part of diet of vast majority of population. Currently, the liquid milk segment accounts for approximately 61% of the overall Indian dairy market, traditional value-added products contribute 34%, and emerging value-added products comprise the remaining 5%. In spite of the vast diversification in the dairy products and the emergence of the value added dairy products, the lactose hydrolyzed market in India is still in the infancy stage.

Even though, the Indian population is most lactose intolerant than many other countries in the world, it is known that 25% of consumers in India are motivated to buy or have started to buy lactose-free drinks (Li et al., 2023a, b). The Mark N Tel Advisors reported that the India lactose free dairy products market is projected to grow at a CAGR of around 7% during the forecast period, i.e., 2022–27 (MarkNTel, 2024). India's largest dairy cooperative, GCMMF, launched India's first branded lactose-free milk in the year 2015. Milky Mist and Raw Pressery also offers similar product in tetra pack with re-closure pack. Milky Mist lactose-free milk pack label displays the content of lactose; galactose and glucose are nil, 2.4 and 2.4%, respectively. Similar label declaration found on other leading manufacturers. Indian Tobacco Company also launched lactose-free milk under the brand of Aashirvaad Svasti 'Easy Digest Milk'. Akshayakalpa also offers lactose-free milk in pouch while Provilac supplies in glass bottle.

2.3 Market Dynamics

2.3.1 Drivers

Lactose intolerance is a common disease; however, it is rare in children younger than 5 years. It is most often seen in adolescents and young adults. Nonetheless, lactose malabsorption is widespread in most of the world. Since the lactose-free dairy products are nutritionally same as that of the regular milk and the growing population of the lactose intolerance and the nutritional awareness has created more demand for the lactose hydrolyzed milk across the world (Bailey et al., 2013). In this aspect, the efforts were beginning in the early 1980's. The first lactose-free milk using lactase in the US was developed at the United States Department of Agriculture (USDA) Agricultural Research Service in 1985 before being commercialized by Lactaid Inc. Valio's technology was applied in 2001 to produce the world's first truly lactose free milk drink, It contains less than 0.01% lactose. Asian populations are genetically predisposed to lactase-deficiency; Monmilk Xinyangdao was certified by the Lactose Intolerance Global Network in 2017, becoming China's first lactose-free dairy brand certified by the international organization. These efforts are all leading lactose-free products to become more mainstream, standardized, and healthier.

In general, lactose hydrolyzed milk market is characterized by innovation, product differentiation, and strategic partnerships to cater the evolving dietary preferences of a lactose intolerance consumes. Increase in the health awareness and wellness concern among the consumers is the key factors that projected to drive the demand and growth of lactose hydrolyzed milk and products. Increase in the self-diagnosis case of lactose intolerance among the consumer is also a supplementary factor of the positive market dynamics. By considering the prevalence of lactose population market segment and potential market opportunities, several dairy industries have initiated intensive research and development on this line. Investing more inputs for the innovative approaches of processing of lactose hydrolysis products, ingredients formulation etc., to improve the quality and variety of lactose hydrolyzed dairy products to cater the need of lactose intolerant population.

2.3.2 *Restraints*

High price of lactose hydrolyzed products is one of the major restraints (Rizzo et al., 2020). Lactose intolerant populations who exclude dairy from their dietary habits are linked to the malnourishment and associated disorders (Bailey et al., 2013). Therefore, by considering the cost component and potential market opportunities, many non-milk alternatives are posing stiff competition to the lactose hydrolyzed products segment. However, several dairy alternatives are either excessively processed to overcome the typical native flavour, or flooded with additives in terms of binders, thickeners, sugar, etc. to match and balance the formulation with respect to the taste consistency and acceptance; therefore. Meeting the regulatory labeling compliance with respect to the level of lactose in low-lactose milk" (shall have less than 1% lactose); and "lactose free milk" (shall have less than 0.1% lactose) or lactose-free dairy products, in which the lactose is hydrolyzed without compromising the quality. Even though, because of the versatility in the lactose hydrolyzed product profile, driven by consumer awareness; still the lactose hydrolyzed products market is increasing marginally as compared to the other products.

2.3.3 *Opportunities*

Due to increasing awareness of lactose intolerance among consumers and demand for healthy dairy options. Lactose hydrolyzed products are not just confined to milk, there are lot many products which can be made lactose free. As a greater number of consumers seeks lactose-free alternatives without compromising taste and nutritional value, the market has expanded to include a diverse range of products such as fresh milk, yogurt, cheese, and ice cream. Gluten-free and low/no/reduced sugar claims in lactose hydrolyzed products help to widen the market base. Lactose hydrolyzed milk products with probiotics, for example, are among the latest trends in the lactose-free products market (Ahn et al., 2023).

2.4 Additional Costs of Lactose Hydrolyzed Dairy Products

Since the production of lactose hydrolyzed milk involves treatment of the regular milk with lactase enzyme, this involves intensive quality assurance to meet the regulatory compliance with respect to the lactose content in the hydrolyzed product. Within the existing facility, apart from the regular milk processing, there is a considerable investment is required towards the energy consumption and maintenance cost to provide the optimal time temperature during the entire hydrolysis reaction. Batch and continuous processing of lactose hydrolyzed has major impact on the cost component to achieve the production efficiency, cleaning, replacement of filters and routine preventive maintenance. Further, cost of enzyme in terms of source and purity, additional storage, packaging and routine quality control aspects pose significant impact on the cost dynamics of the lactose hydrolyzed products.

2.5 Future Growth of Lactose-Free Products in Market

Lactose hydrolyzed dairy products are nutritionally healthier than alternative lactose-free non-dairy products. Lactose hydrolyzed milk is a rich source of high-quality proteins, superior amino acid profile, calcium, phosphorus, vitamin B12, riboflavin and Vitamin D. The surge in health benefits offered by lactose hydrolyzed dairy products has resulted from increased innovation in the lactose hydrolyzed dairy market, which represents important growth opportunities for the lactose-free products market.

Ultra-heat-treated lactose-free milk is the major option found in many countries, because of this the sales of the lactose hydrolyzed dairy product is increasing marginally. However, the increasing number of launches in the chilled section is indicating the move to mainstream where it is easier for consumers to find it. Lactose hydrolyzed dairy products offers great opportunity to the dairy industries to do the value addition in the milk and milk products. In addition to this, cost saving can be achieved in the flavored milk, milk shakes, and traditional sweets by reducing the sugar level as lactose hydrolysis enhances the sweeter taste (Adhikari et al., 2010). The eating habits of consumers are rapidly evolving. People are more on-the-go, have a hectic schedule, and their eating habits are more erratic. Lactose hydrolyzed product manufacturers are constantly adapting to satisfy the rising demand for lactose-free products from today's consumers. The latest trends in the lactose hydrolyzed dairy product market include products that are relatively compact and more convenient and offerings a range of health benefits.

2.6 Conclusion

Lactose hydrolyzed dairy products category has captured a niche market because of the growing health benefit to all consumers and in countries where most people are lactose tolerant. Lactose hydrolyzed product manufacturers are adapting strategy to satisfy the rising demand for lactose hydrolyzed products from today's consumers. Since the time of commercial production of first lactose free milk to till date huge diversification is perceived to cater the need of lactose intolerant population. Lactose hydrolyzed dairy market is progressively expanding all over the world. At the moment, North America and Europe are the major market player; Asia-Pacific market is expected to create huge market demand by considering the prevalence of lactose intolerant population. Consumer awareness on the health benefits offered by lactose hydrolyzed dairy products and the research and development innovations offered by the dairy industries in the lactose hydrolyzed dairy market plays important role in creating the sustained growth opportunities of the lactose hydrolyzed dairy products market.

Disclaimer Usage of the Trade name, manufacturers details cited in this manuscript is for the academic purpose, not from the commercial point of view.

References

Adhikari, K., Dooley, L. M., Chambers, E., & Bhumiratana, N. (2010). Sensory characteristics of commercial lactose-free milks manufactured in the United States. *LWT—Food Science and Technology, 43*, 113–118.

Ahn, S., Kim, M. S., Park, D. G., Han, B. K., & Kim, Y. J. (2023). Effects of probiotics administration on lactose intolerance in adulthood: A meta-analysis. *Journal of Dairy Science, 106*(7), 4489–4501. https://doi.org/10.3168/jds.2022-22762

Allied Market research. (2022). *Lactose free dairy market*. Accessed at https://www.alliedmarketresearch.com/lactose-free-dairy-market-A18453

Bailey, R. K., Fileti, C. P., Keith, J., Tropez-Sims, S., Price, W., & Allison-Ottey, S. D. (2013). Lactose intolerance and health disparities among African Americans and Hispanic Americans: An updated consensus statement. *Journal of the National Medical Association, 105*(2), 112–127.

BIS Research. (2024). Global lactose-free dairy products market. *A Global and Regional Analysis*, 2023–2033. Accessed at https://bisresearch.com/industry-report/lactose-free-dairy-products-market.html

Data bridge market research. (2022). *Global lactose and derivatives market- Industry trends and forecast to 2029*. Accessed at https://www.databridgemarketresearch.com/reports/global-lactose-and-derivatives-market

Dekker, P. J. T., Koenders, D., & Bruins, M. J. (2019). Lactose-free dairy products: Market developments, production, nutrition and health benefits. *Nutrients, 11*(3), 551. https://doi.org/10.3390/nu11030551

Euromonitor Database. (2019). Available online: https://www.euromonitor.com.

GOI. (2024). *World milk day*. Press Information Bureau, Government of India. Accessed at https://pib.gov.in/PressNoteDetails.aspx?NoteId=151889&ModuleId=3

Hernandez, A. J., Truong, T., Barbano, D. M., & Drake, M. A. (2024). Milk beverage base with lactose removed with ultrafiltration: Effect of fat and protein concentration on sensory and physical properties. *Journal of Dairy Science, 107*(1), 169–183.

Kamrath, C., Bidkar, S., & Bröring, S. (2019). Is food involvement in purchasing decisions always low? A consumer study from Germany. *Pharma Nutrition, 9*, 100157.

Lammers, H. B. (1991). The effect of free samples on immediate consumer purchase. *Journal of Consumer Marketing, 8*(2), 31–37.

Laxmana, N. N., Nath, B. S., Prasad, R., Vimala, B. S. M., & Nandini, H. (2018). Food labelling: An overview. *Indian Journal of Dairy and Biosciences (IJDBS), 29*(1–8), S68–S71.

Li, A., Zheng, J., Han, X., Yang, S., Cheng, S., Zhao, J., & Lu, Y. (2023a). Advances in low-lactose/lactose-free dairy products and their production. *Food, 12*(13), 2553.

Li, A., Zheng, J., Han, X., Yang, S., Cheng, S., Zhao, J., Zhou, W., & Lu, Y. (2023b). Advances in low-lactose/lactose-free dairy products and their production. *Food, 12*, 2553. https://doi.org/10.3390/foods12132553

Malik, T. F., & Panuganti, K. K. (2023). *Lactose intolerance*. StatPearls Publishing.

Market Research Future. (2024). *Lactose-free products dairy market research report*. https://www.marketresearchfuture.com/reports/lactose-free-dairy-products-market-6434

Markets and Markets. (2022). *Lactase market*. Accessed at https://www.marketsandmarkets.com/Market-Reports/lactase-market-125332780.html

MarkNTel. (2024). *India lactose-free dairy products market research report*. Accessed at https://www.marknteladvisors.com/query/request-sample/india-lactose-free-dairy-products-market.html

Misselwitz, B., Butter, M., Verbeke, K., & Fox, M. R. (2019). Update on lactose malabsorption and intolerance: Pathogenesis, diagnosis and clinical management. *Gut, 68*(11), 2080–2091.

Morrissey, P. A. (1985). Lactose: Chemical and physicochemical properties. In *Developments in dairy chemistry—3: Lactose and minor constituents* (pp. 1–34). Springer Netherlands.

Randell, A. W. (2010). The codex Alimentarius and food labelling: Delivering consumer protection. In *Innovations in food labelling* (pp. 5–16). Woodhead Publishing.

Rizzo, P. V., Harwood, W. S., & Drake, M. A. (2020). Consumer desires and perceptions of lactose-free milk. *Journal of Dairy Science, 103*, 6950–6966.

Santeramo, F. G., Carlucci, D., De Devitiis, B., Seccia, A., Stasi, A., Viscecchia, R., & Nardone, G. (2018). Emerging trends in European food, diets and food industry. *Food Research International, 104*, 39–47.

Ségurel, L., & Bon, C. (2017). On the evolution of lactase persistence in humans. *Annual Review of Genomics and Human Genetics, 18*(1), 297–319.

Silanikove, N., Leitner, G., & Merin, U. (2015). The interrelationships between lactose intolerance and the modern dairy industry: Global perspectives in evolutional and historical backgrounds. *Nutrients, 7*, 7312–7331. https://doi.org/10.3390/nu7095340

Suchy, F. J., Brannon, P. M., Carpenter, T. O., Fernandez, J. R., Gilsanz, V., Gould, J. B., & Wolf, M. A. (2010). NIH consensus development conference statement: Lactose intolerance and health. *NIH Consensus and State-of-the-Science Statements, 27*(2), 1–27.

Vardhanabhuti, B., & Wang, Y. (2022). Low lactose milk products. In *Encyclopedia of dairy sciences* (3rd ed., pp. 423–429).

Chapter 3
β-Galactosidases Enzymes for Lactose Hydrolysis

C. T. Manoj Kumar, M. H. Sathish Kumar, S. Supreetha, and Deepak

3.1 Introduction

The β-galactosidase is a glycoside hydrolase enzyme that hydrolyses the lactose into glucose and galactose. It is commonly used for the hydrolysis of lactose in milk to serve it for individuals suffering lactose intolerance. It is also used for lactose hydrolysis in condensed milk to improve the texture of the product by escaping the crystallization and in whey to assist to produce ethanol. In addition to that, it is used for pharmaceutical and biosensor based analytical applications. In addition to lactose hydrolysis, it has exhibits transgalactosylation activity to produce galactooligosaccharides (GOS) from lactose. Basically, GOS is a dietary fiber which is widely known for its prebiotic property and various physiological activities. It is commonly used in infant formulas due to its physiological activity akin to dietary fibres present in mother's milk. The global β-galactosidase market size is expected to grow at a CAGR of 4.5% from 2023 to reach US$ 1360.9 million by 2029 (Valuates Reports, 2024). The structure of the β-galactosidase is complex in nature (tetramer) and it is very well reported along with its mode of action over substrate. The β-galactosidases can be obtained from various sources such as plants, microorganisms and animals. It is commercially produced from various bacteria, yeast and molds due to easy and low-cost maintenance. The microorganisms such as *Escherichia coli*, *Bacillus* Sp., *Aspergillus oryzae*, *Kluyveromyces Sp.*, *Bifidobacterium Sp.*, etc. have been well explored to produce β-galactosidase. The recombinant form of β-galactosidase production is very well studied and successfully adopted for commercial production to reduce cost of production and to improve activity and stability of the enzyme. However, the source for the β-galactosidase production must be of food grade for food processing applications. The analytical and pharmaceutical applications are

C. T. Manoj Kumar (✉) · M. H. Sathish Kumar · S. Supreetha · Deepak
Dairy Technology Section, Southern Regional Station, ICAR-National Dairy Research Institute, Bengaluru, Karnataka, India

P. S. Rao et al. (eds.), *Lactose Hydrolysis in Dairy Products*,
https://doi.org/10.1007/978-3-031-78207-7_3

dominated by *E. coli* β-galactosidase, while food processing applications are dominated by *A. oryzae*, *K. lactis.*, *B. circulans*, etc. β-galactosidases. The β-galactosidase can be bioengineered using various genome editing techniques to enhance its stability to extreme temperature, pH and reaction products and to widen its substrate specificity to enhance its potential. The stability and potential applications of β-galactosidase can be increased using immobilization techniques. The immobilized β-galactosidase can be used multiple times which reduces cost of production and increases purity of the final product.

3.2 Classification of β-Galactosidase

The enzymes that exhibit β-galactosidases activity (such as lactose hydrolysis) are classified under glycoside hydrolase (GH) families such as GH1, GH2, GH35, GH42, GH59 and GH147 (Lu et al., 2020). Among all these, the maximum numbers of identified β-galactosidases are belonging to GH1, GH35 and GH42 families. The GH2 family contains β-galactosidases mainly produced from bacteria and yeast (Jacobson et al., 1994; Lu & Xiao, 2017). The *Escherichia coli* based β-galactosidase is most commonly studied under this family. The GH35 β-galactosidases belong to bacteria, fungi, plants and animals. The GH42 β-galactosidases are mainly obtained from bacteria, especially thermophiles. The GH1 contains very few β-galactosidase obtained from archaea. Unlike other β-galactosidases, it can hydrolyze glycosyl moieties other than galactose.

3.3 Structure of β-Galactosidase Enzymes

The three-dimensional structure of β-galactosidase was elucidated using orthorhombic crystal form with a single tetramer in the asymmetric unit. The β-galactosidase contains four polypeptide chains (tetramer) referred as monomer A-D. Each monomer made up of 1023 amino acids and divided into five domains which are respectively colored blue (Domain 1), green (Domain 2), yellow (Domain 3), cyan (Domain 4) and red (Domain 5) (Juers et al., 2000; Matthews, 2005). The enzyme contains four catalytic sites that show no cooperativity or allosteric effectors (Juers et al., 2000). Domain 3 has an α/β or 'TIM' barrel structure with the active site located at the C-terminal end of the barrel by forming deep bottom (Matthews, 2005). The formation of active site is contributed by components of different segments of polypeptide chain. The loops or components required for active site may come from different domains of same monomer or from other monomers. The domain 2 loop of Monomer A required to complete the active site of Monomer D. Similarly, the active site of monomer A works with the support of a loop from Monomer D. Similar exchange can also be seen between Monomer B and

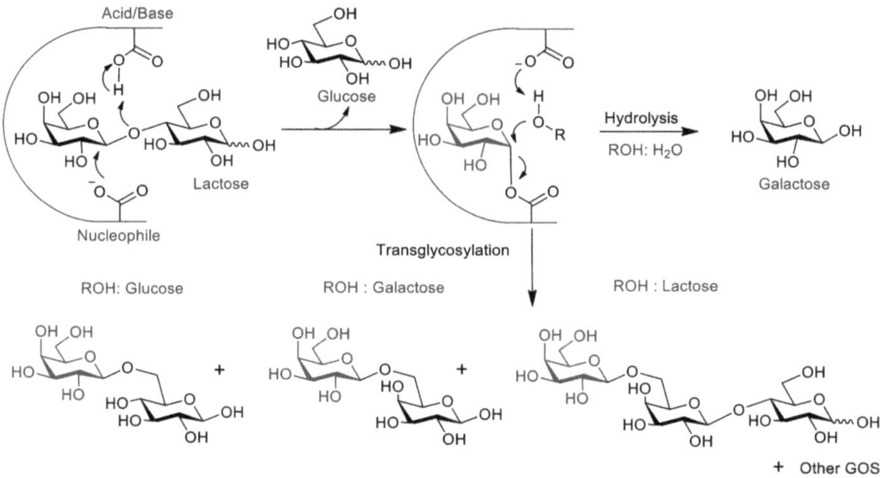

Fig. 3.1 Lactose hydrolysis and transgalactosylation activities of β-galactosidase. (Lu et al., 2020)

C. However, these active sites are well separated and act independently. The active sites also contain metal ions such as Mg^{2+} and Na^{+} which are crucial for catalytic activity of β-galactosidase (Wallenfels & Weil, 1972).

In tetramer polypeptide chain, three distinct interfaces are formed between different pairs of monomers. These interfaces are grouped by using three mutually-perpendicular two-fold axes of symmetry. It can be seen from Fig. 3.1 that the 'long' interface, formed by the horizontal two-fold axis by relating the A with B and C with D monomer. The 'activating' interface, formed by the vertical two-fold axis by relating A with D, and B with C. The third interface relates A with C and B with D (Juers et al., 2000).

3.4 Source of β-Galactosidase Enzymes

The β-galactosidase is naturally present in various sources such as mammals, microorganisms (bacteria, yeast and fungi) and plants (Panesar et al., 2010). For the commercial production, it is usually extracted from various native and recombinant microorganisms. Because, they are easy to cultivate and extract β-galactosidase, inexpensive, mass production is possible, possible for genome editing and some of them carries generally regarded as safe (GRAS) status (suitable for food applications). However, its structure, optimum operating conditions (temperature and pH), stability and substrate specificity may vary with the source. For example, the optimum operating pH of fungal β-galactosidase is 2.5 to 5.4, however, it is 6.0–7.0 pH for yeast and bacterial β-galactosidase. Hence, the fungal β-galactosidase is preferred for lactose hydrolysis in acidic whey. Similarly, β-galactosidase from several sources shows better transgalactosylation activity due to substrate specificity.

3.4.1 β-Galactosidase Extracted from Bacterial Origin

Bacterial β-galactosidases have certain advantages such as high activity (U/mL), ease in fermentation and more stable enzyme. The bacteria such as *Escherichia coli, Bifidobacterium, Lactobacillus, Bacillus* have been used for the production of β-galactosidase (Table 3.1). The most commonly studied β-galactosidase is from *E. coli* but not used widely due to lack of "generally regarded as safe (GRAS)" status.

3.4.2 β-Galactosidase Extracted from Yeast and Fungi

Among yeasts, β-galactosidase most found in *Saccharomyces fragilis* and *Kluyveromyces lactis* (Table 3.2). The *Kluyveromyces lactis* is commonly found in the milk and milk products and is one of the important sources of β-galactosidase. Mold β-galactosidases are highly stable to both acidic pH (2.5–5.4) and high temperatures (40–65 °C) making them more useful in acidic system applications like acidic whey. *Aspergillus oryzae* and *Aspergillus oryzae* have great commercial importance and widely used fungal β-galactosidase. The microorganisms such as *Aspergillus species, Kluyveromyces lactis* and *Saccharomyces fragilis* are GRAS. In contrast to yeast enzyme, the fungi β-galactosidase is active without cofactors such as Mn^{2+} and Na^{+} (Jurado et al., 2002). However, the presence of Ca^{2+} and heavy metals is detrimental to all β-galactosidases.

Table 3.1 β-galactosidases from bacterial sources

Source microorganism	Use	References
E. coli	Immobilization	Crescimbeni et al. (2010)
Bacillus circulans	Lactose hydrolysis	Das et al. (2015)
Lactobacillus plantarum	Lactose hydrolysis	Selvarajan and Mohanasrinivasan (2015)
Thermotoga maritima	Lactose hydrolysis / immobilization / GOS synthesis	Marín-Navarro et al. (2014)
B. circulans	GOS synthesis / immobilization	Urrutia et al. (2014)
Lactobacillus leichmannii	GOS synthesis	Ji et al. (2021)
Lactococcus lactis	GOS synthesis	Yu and O'sullivan (2014)
Bifidobacterium longum	GOS production	Hsu et al. (2007)
B. longum CCRC 15708, *B. longum* B6 and *B. infantis* CCRC 14633	Lactose hydrolysis / GOS synthesis	Hsu et al. (2005)

Table 3.2 β-galactosidases from various yeast and mold sources

Source	Applications	References
Aspergillus oryzae	Lactose hydrolysis/Immobilization	Eldin et al. (2005)
A. oryzae	Lactose hydrolysis/Immobilization	Haider and Husain (2008)
A. oryzae	Lactose hydrolysis/Immobilization	Ansari et al. (2012)
Kluyveromyces lactis	Immobilization	Cabuk et al. (2014)
K. fragilis	Lactose hydrolysis	Matioli et al. (2003)
K. fragilis	Lactose hydrolysis/Immobilization	Roy and Gupta (2003)
K. marxianus	Ethanol production	Gupte and Nair (2010)

3.5 Expression Hosts for β-Galactosidases

Recombinant protein expression involves isolation of a gene of interest from an organism (called native organism) and to express it in another organism (called host organism) (Raguin et al., 2023). The purpose of recombinant protein is to extract and purify quickly large amounts of a target protein (enzyme) expressed in a selected host for bulk production. The method helps to obtain higher yields of pure, stable and highly functional enzyme compared to extraction from native organism. This approach can be used for large scale production of β-galactosidase having distinguished characteristics such as pH (acidic condition) and temperature (low and high temperature) tolerance and better transgalactosylation activity. This expands the potential large-scale applications of the β-galactosidase and makes it economically viable for industrial production. The host organism must be easy and inexpensive to cultivate and carry the 'generally recognized as safe (GRAS)' or 'qualified presumption of safety (QPS)' status for food applications (Nguyen et al., 2015). Several microorganisms have been used as a host for β-galactosidase production such as *Escherichia coli* (*E. Coli*), *Bacillus subtilis*, *Saccharomyces cerevisiae*, *Kluyveromyces lactis*, *Pichia pastoris*, *Lactiplantibacillus plantarum*, *Latilactobacillus sakei*, *Lactococcus lactis*, etc. (Oliveira et al., 2011).

3.5.1 Escherichia coli

E. coli is the most extensively explored expression host for the production of β-galactosidase. It is popularly used host for industrial production of pharmaceutical proteins due to most accessible genetic information, feasible genetic manipulation, high protein yield and cost effective (Westers et al., 2004; Movahedpour et al., 2022). It can produce exceptionally higher yield of pure, active and stable β-galactosidase than other host organisms. The higher activity of recombinant β-galactosidase could be due to higher purity. The fusion-based purification method such as His (six-histidine) tag is commonly used for purification of β-galactosidase from *E. coli*. Several literatures are reported on production of recombinant β-galactosidase using *E. coli* (Table 3.3). Nguyen et al. (2007) reported that heterodimeric β-galactosidase of

Table 3.3 Recombinant β-galactosidases produced in various expression hosts

Source of the enzyme	Expression host	Culture volume and β-galactosidase purification yield	References
Bacillus circulans	*E. coli* BL21(DE3)	3 L, 338 U ml^{-1}	Duan et al. (2017)
Pseudoalteromonas haloplanktis TAE79 *E. coli* *Bacillus stearothermophilus*	*Bacillus subtilis*	43.1 U mg^{-1} 627.5 U mg^{-1} 61.0 U mg^{-1}	Welsch et al. (2012)
Bacillus aryabhattai	*Bacillus subtilis*	17.41 U/mL	Duan and Luan (2023)

Lactobacillus reuteri L103 is cloned and overexpressed in *E. coli* using an expression system based on the T7 RNA polymerase promoter. The heterologous expression produced multiple fold β-galactosidase (83 kU/L) with a specific activity of 180 U/mg than natural source (2.5 kU/L) with a specific activity 158 U/mg. The minor change in recombinant enzyme's properties such as stability, optimum temperature and pH and activity was observed. The higher substrate affinity and hydrolytic activity of recombinant enzyme was observed when lactose was used as substrate. The recombinant enzyme was purified with single step whereas natural source enzyme was purified with three consecutive steps. Similar results reported for β-galactosidase from *Thermus* sp. strain and *Alicyclobacillus acidocaldarius* expressed in *E. coli* (Di Lauro et al., 2008; Pessela et al., 2003). However, the recombinant β-galactosidase expressed by *E. coli* is preferred for pharmaceutical, analytical and biosensor applications but not for commercial food processing applications as it is not considered as safe (Panesar et al., 2006).

3.5.2 Bacillus

Bacillus genus-based bacteria have been used for the expression of enzymes suitable for industrial application because of their ability to secrete a large number of particular proteins under moderate or low incubation temperatures (Nijland & Kuipers, 2008). The bacteria belong to Bacilli genus such as *Bacillus subtilis, Bacillus licheniformis*, and *Bacillus amyloliquefaciens* have been explored for the expression of β-galactosidase from various microorganisms. Among them, the *B. subtilis* is widely used for the expression (mostly homologous expression) of industrial enzymes such as alkaline proteases, amylases, etc. In contrast to *E. coli*, the major advantage with *B. subtilis* is that it has generally recognized as safe (GRAS) status so that it can be used for the production of enzymes used for food applications (Martinez et al., 1997; Movahedpour et al., 2022). In addition to that, it has naturally high secretory capacity and unlike *E. coli*, it exports proteins directly into the extracellular medium (Simonen & Palva, 1993). Further, it is easy to cultivate and inexpensive, expressed high secretion capacity and adoptable to downstream processing (Movahedpour et al., 2022). Despite these benefits, its application is limited to the bulk production of industrial enzyme rather than analytical and pharmaceutical enzymes. The draw backs for these limitations could be due to lack of suitable expression vectors, plasmid instability, and presence of proteases and occurrence of malfolded proteins (Westers et al., 2004). In addition to that, *B. subtilis* secretion pathway suffers from improper translocase targeting, secreted protein degradation and inappropriate folding (Westers et al., 2004). However, further studies are required to understand the mechanisms and regulation of pathways to overcome these drawbacks.

3.5.3 Saccharomyces

The flocculating type yeast organism *Saccharomyces cerevisiae* is commonly evaluated for expression of β-galactosidase from different microorganisms (Domingues et al., 1999, 2000, 2004). The flocculating *S. cerevisiae* has extracellular enzyme production system and maintains high cell density levels during sedimentation in bioreactor. The flocculated high cell density offers advantages such as higher recombinant enzyme production at continuous mode with minimal risk of contamination. It also simplifies downstream purification of enzyme due to sedimentation (Domingues et al., 2004; Oliveira et al., 2011). The basic purpose of expression of β-galactosidase in *S. cerevisiae* was to enable the lactose hydrolysis to produce ethanol as it lacks in lactose permease system. During an attempt, the flocculating *S. cerevisiae* NCYC869-A3 modified with the pVK1.1 vector successfully produced 2X times more β-galactosidase activity in less lactose medium (Domingues et al., 2002). With success of recombinant production of β-galactosidase with *S. cerevisiae* allowed further studies on large scale production using lactose solution (2 L) and cheese whey (10 L) (Domingues et al., 2002, 2004). The recombinant method showed 21-fold increase in β-galactosidase in cheese whey. This achievement marked successful application of recombinant method in ethanol production directly from cheese whey. The continuous production of recombinant β-galactosidase was demonstrated 6 L bioreactor (Domingues et al., 2005). The recombinant β-galactosidase productivity was increased by 11-fold compared to the batch system. However, plasmid stability was low at higher dilution rates leads to decrease in β-galactosidase productivity. This limitation was sorted with intrusion of lacA expression cassette in the δ-sequences of the genome of the *S. cerevisiae* NCYC869 (Oliveira et al., 2007).

3.6 Engineering of β-Galactosidases

The characteristics of β-galactosidase can be modified by using several techniques such as clustered regularly interspaced palindromic repeats-associated (Cas) proteins (CRISPR/Cas-9), transcription activator-like effector nucleases (TALENs), zinc-finger nucleases (ZFNs), site-directed mutagenesis, truncation, site-saturation mutagenesis, random mutagenesis, DNA shuffling and monobody modifications. These techniques usually involve genome editing to improve the functional properties of enzymes such as transgalactosylation activity, broader substrate specificity, stability to extreme environmental factors (temperature, pH, etc.), and susceptibility to reaction product products (Movahedpour et al., 2022). The engineered β-galactosidase (also called as mutants) can be produced through heterologous expression using various microorganisms. Several studies have been conducted to improve the efficiency of galactooligosaccharides (GOS) production, broadening the specificity, reduce reaction products susceptibility and stability of β-galactosidase through genome editing are discussed here below.

Transgalactosylation activity of β-galactosidase is to synthesize the GOS from lactose. This is the intermediate step occurs during hydrolysis of lactose. The enzyme forms complex with galactose and transfer it to another nucleophile acceptor. However, the production of GOS affected by stability and poor transgalactosylation activity of β-galactosidase. Few studies have been carried out to improve transgalactosylation activity of β-galactosidase through genome editing. Wu et al. (2013) performed site directed mutagenesis for β-galactosidase from *Sulfolobus solfataricus* P2 and obtained two mutants F359Q and F441Y. Both mutants produced 58.3 and 61.7% GOS, respectively while the native enzyme produced 50.9%. Placier et al. (2009) applied saturated mutagenesis applied for β- galactosidase (BgaB) from *Geobacillus stearothermophilus* KVE39 and generated two mutants GP21 (BgaB R109K) and GP637.2 (BgaB R109V). Both mutants produced 11.5 and 21% GOS, respectively while native enzyme achieved 2% GOS. Similarly, Jorgensen et al. (2001) used truncated mutation technique to transform β-galactosidase BIF3 (188 kDa, 1752 amino acids residues) from *Bifidobacterium bifidum* DSM20215. The truncated β-galactosidase managed to utilize 90% of lactose and maintains 9:1 transgalactosylation to hydrolysis ratio.

The enzymes are strictly specific to particular linkage of the substrate for certain reactions. Similarly, β-galactosidase is specific to β-D-galactosyl linkage of the substrate (galactose moiety of the lactose). However, the specificity of the enzyme can be altered through genome editing. Several studies reported that genome editing (such as DNA shuffling) of β-galactosidase modified its specificity similar to β-fucosidase and β-glucuronidase (Zhang et al., 1997; Xiong et al., 2007). The shift in the specificity of the β-galactosidase increases its scope in various applications. The genome editing also reported for broadening the glycosyl acceptor specificity of the β-galactosidase. Generally, the β-galactosidase catalyzes the formation of O-glycosidic bonds between galactose and alcoholic hydroxyl groups of the acceptors. However, glycosylation of phenolic hydroxyl or glycosides having other than O-linkage is difficult due to poor nucleophilicity (Kim et al., 2010). Studies reported on glycosylation of catechol, hydroquinone, phenol, pyrogallol and caffeic acid using mutant β-galactosidase (W980F) (Lu et al., 2015a, b).

The stability of the β-galactosidase against extreme temperature, pH and reaction products is important as it will face them during dynamic industrial production processes. Several reports are available over improved β-galactosidase stability post genome editing process. Ishikawa et al. (2015) reported that K166P, G307P, A833P mutants were generated from BgaD-D β-galactosidase using site-directed mutagenesis technique had better thermal stability at 60 °C than native enzyme. Similarly, Rico-Díaz et al. (2017) reported that the mutated β-galactosidase (R116C, T270C, G818C) developed by incorporating cysteine in β-galactosidase from *K. lactis* had 6.8 times higher half-life at 45 °C. The mutants showed better GOS production ability than native enzyme. The β-galactosidase is highly susceptible for the reaction product (galactose) released during lactose hydrolysis. The galactose competitively inhibits the lactose binding to the active site of β-galactosidase. Several galactose in-sensitive mutants were developed from β-galactosidase of *Aspergillus niger* and *Aspergillus candidus* (Hu et al., 2010; Zhang et al., 2018).

3.6.1 Hydrolytic and Transgalactosylation Activity

β-galactosidase catalyzes the hydrolysis of lactose into glucose and galactose. The catalysis is done through two step double-displacement mechanism (retaining mechanism) and it is executed using two glutamic or aspartic acid residues of β-galactosidase (Lu et al., 2020) (Fig. 3.2). The one glutamic or aspartyl residue acts as nucleophile (glutamate 551 (Glu_{551}), Glu_{537}) and another one as acid/base catalyst or proton donor (Glu_{482}, Glu_{461}) (Juers et al., 2001; Zhou & Chen, 2001; Lu et al., 2020). In the first step, the nucleophile covalently binds to the anomeric center of galactose moiety (forms intermediate galactosyl-enzyme complex), whereas the acid-base residue/proton donor helps (acid assistance) for release of glucose moiety. The sodium ion may also help in binding by directly ligating the galactose moiety. In the second step, the acid base residue activates the acceptor molecule by acting as a general base. The activated molecule binds to the anomeric carbon center of galactosyl-enzyme intermediate without affecting the anomeric configuration of substrate. Hydrolysis takes place if water binds to the galactosyl-enzyme intermediate as an acceptor molecule and the galactose is released from the galactosyl-enzyme

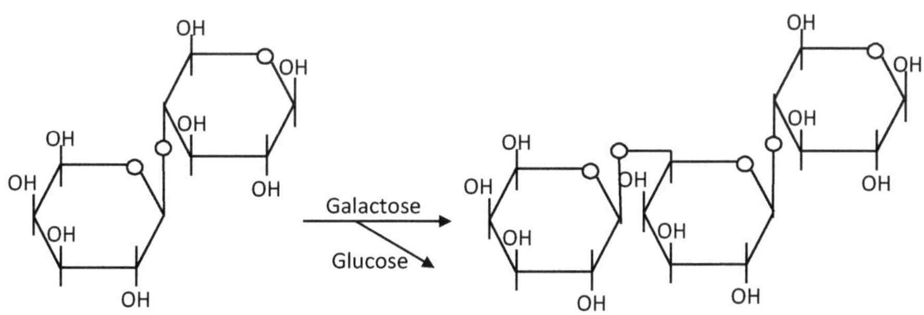

Fig. 3.2 Trisaccharide GOS (β-D-Galactose-(1 → 6)-β-D-Galactose-(1 → 4)-D-Glucose) production using lactose as acceptor

Table 3.4 List of oligosaccharide structures identified in GOS preparations

GOS component	Chemical structure
Disaccharides	β-D-Gal (1 → 6)-D-Glc
	β -D-Gal (1 → 6)-D-Gal
	β -D-Gal (1 → 3)-D-Glc
	β -D-Gal (1 → 2)-D-Glc
	β -D-Gal (1 → 3)-D-Gal
Trisaccharides	β -D-Gal (1 → 6)- β -D-Gal (1 → 6)-D-Glc
	β -D-Gal (1 → 6)- β -D-Gal (1 → 4)-D-Glc
	β -D-Gal (1 → 6)- β -D-Gal (1 → 6)-D-Gal
	β -D-Gal (1 → 3)- β -D-Gal (1 → 4)-D-Glc
	β -D-Gal (1 → 4)- β -D-Gal (1 → 4)-D-Glc
Tetrasaccharides	β -D-Gal (1 → 6)- β -D-Gal (1 → 6)-D-Gal (1 → 4)-D-Glc
	β -D-Gal (1 → 6)- β -D-Gal (1 → 3)-D-Gal (1 → 4)-D-Glc
	β -D-Gal (1 → 3)- β -D-Gal (1 → 6)-D-Gal (1 → 4)-D-Glc
Pentasaccharides	β -D-Gal (1 → 6)- β -D-Gal (1 → 6)-D-Gal (1 → 6)-D-Gal (1 → 4)-D-Glc

intermediate. Transgalactosylation takes place if another saccharide (mono, di, tri-saccharides, etc.) binds to the galactosyl-enzyme intermediate complex as an acceptor and new galactooligosaccharides will form (Lu et al., 2020). Juers et al. (2001) reported that galactose moiety will move deep into the active site pocket of β-galactosidase during the reaction. During this process, a conformational change can be seen with those ligands which bind to β-galactosidase. The residues within loop 794–804 may advance up to 10 Å closer to the site of binding. However, it can be inhibited by the binding of additional ligands.

The lactose rich media favours the formation of more and higher galactooligosaccharides. GOS is synthesized by linking galactose to glucose with various types of glycosidic bonds, such as β-1 → 2, β-1 → 3, β-1 → 4, and to galactose through bonds β-1 → 6 and β-1 → 3. The intermediate galactosyl-enzyme complex may also bind to another lactose with 1 → 6 linkage (Fig. 3.1) to form tri GOS. The β-galactosidase enzyme synthesizes different GOS products (based on number of monosaccharides attached) namely di, tri, tetra and higher saccharides (Table 3.4).

Quantitatively, the amount of the different GOS products present appears to follow the order: di- > tri- > tetra- > higher- saccharides and the linkages synthesized are predominantly β-(1 → 6) > β-(1 → 3) and β-(1 → 2). Trisaccharides, especially galactosyl 1 → 6 lactose, can be identified at most lactose levels. Tetra- and higher saccharides have been reported only when using much higher initial lactose levels.

3.7 Commercial Enzyme in Lactose Hydrolysis

As demand for β-galactosidase increasing in the food processing applications, various commercial β-galactosidase producers can be seen in the market. Commercially available β-galactosidases are summarized in the Table 3.5.

Table 3.5 Commercially available β-galactosidase

Trade name	Source	Units	Company
Lactozym® Pure 6500 L Lactozym 2600 L	*K. fragilis*	6500 LAU/g 2600 LAU/g	Novozymes, Denmark
Maxilact® NEXT	*K. lactis*	NA	DSM food specialties, Netherlands
Maxilact LX 5000	*K. lactis*	5000 NLU/g	DSM food specialties, Netherlands
Maxilact® LGi 2000	*K. lactis*	2000 NLU/g	DSM food specialties, Netherlands
MAXILACT® LAGX 1000	*K. lactis*	NA	DSM food specialties, Netherlands
Maxilact® LAGX 2000	*K. lactis*	NA	DSM food specialties, Netherlands
Biolactasa-NTL conc X2 Biolactase F Conc	*B. Circulans* *A. oryzae*	553 U/mL	BIOCON, Spain
Lactase	*A. oryzae*	100,000 U/g	Anthem biosciences, India
Lactase	*E. coli*	500 U/mg	Sisco research laboratories, India
GODO-YNL2 lactase	*K. lactis*	5000 NLU/g	DuPont Danisco, Copenhagen, Denmark
Validase fungal lactase concentrate Validase yeast lactase	*A. oryzae* *K. lactis*	50,000 U/g	Valley research, Inc., South Bend, USA
Enzeco fungal lactase	*A. oryzae*	152,361 ± 4755 IU/g protein	Enzyme development corporation, New York, USA
Lactase	*A. oryzae*	100,000 ALU/g	Bio-Cat Inc., USA
Lactomax pure Lactomax pure 200 Lactomax F30	*K. Lactis* *A. oryzae*	NA	Prozyn, Sao Paulo, Brazil
Biolactase STR20	NA	NA	Kerry bioscience, Ireland
Astrolact LX2000 Astrolact LX5000	*K. Lactis*	NA	Calza Clemente, Italy
NOLA™ Fit 5500	*B. bifidum*	NA	Kerry bioscience, Ireland
Ha-lactase™ 5200	*K. lactis*	8040 U/g	Kerry bioscience, Ireland

NA Not Available

3.8 Conclusion

The β-galactosidase is a valuable tool which is useful for lactose free milk and milk products preparation, synthesis of galactooligosaccharides, pharmaceutical and biosensors applications. These enzymes are readily available, inexpensive, easy to handle, and flexible to simple substrates without requirement of cofactors. It is worth noting that the capability of β-galactosidases is continuously expanding nowadays. Numerous current studies are focusing on β-galactoidases with high transglycosylation activity, which help to increase oligosaccharide production and reduce large-scale production costs. Various vectors are developed for better expression of recombinant enzyme. Molecular engineering of β-galactosidases greatly improves transglycosylation activity and/or substrate tolerance, and extends the application to synthesize unusual and challenging structures. New genome editing techniques such as Crispr/cas-9 are can be used for further development of recombinant enzyme production and bioengineering of enzyme.

References

Ansari, S. A., Satar, R., Alam, F., Alqahtani, M. H., Chaudhary, A. G., Naseer, M. I., et al. (2012). Cost effective surface functionalization of silver nanoparticles for high yield immobilization of aspergillus oryzae β-galactosidase and its application in lactose hydrolysis. *Process Biochemistry, 47*(12), 2427–2433.

Cabuk, B., Tari, C., & Harsa, S. T. (2014). β-Galactosidase immobilization on chitosan-hydroxyapatite complex: Effects of immobilization conditions. *Journal of Nutritional, Health and Food Engineering, 1*, 13–23.

Crescimbeni, M. C., Nolan, V., Clop, P. D., Marín, G. N., & Perillo, M. A. (2010). Activity modulation and reusability of β-d-galactosidase confined in sol–gel derived porous silicate glass. *Colloids and Surfaces B: Biointerfaces, 76*(2), 387–396.

Das, B., Roy, A. P., Bhattacharjee, S., Chakraborty, S., & Bhattacharjee, C. (2015). Lactose hydrolysis by β-galactosidase enzyme: Optimization using response surface methodology. *Ecotoxicology and Environmental Safety, 121*, 244–252.

Di Lauro, B., Strazzulli, A., Perugino, G., La Cara, F., Bedini, E., Corsaro, M. M., et al. (2008). Isolation and characterization of a new family 42 β-galactosidase from the thermoacidophilic bacterium Alicyclobacillus acidocaldarius: Identification of the active site residues. *Biochimica et Biophysica Acta (BBA)-Proteins and Proteomics, 1784*(2), 292–301.

Domingues, L., Teixeira, J. A., & Lima, N. (1999). Construction of a flocculent Saccharomyces cerevisiae fermenting lactose. *Applied Microbiology and Biotechnology, 51*, 621–626.

Domingues, L., Onnela, M. L., Teixeira, J. A., Lima, N., & Penttila, M. (2000). Construction of a flocculent brewer's yeast strain secreting aspergillus Niger β-galactosidase. *Applied Microbiology and Biotechnology, 54*, 97–103.

Domingues, L., Teixeira, J., Penttila, M., & Lima, N. (2002). Construction of a flocculent Saccharomyces cerevisiae strain secreting high levels of aspergillus Niger β-galactosidase. *Applied Microbiology and Biotechnology, 58*, 645–650.

Domingues, L., Oliveira, C., Castro, I., Lima, N., & Teixeira, J. A. (2004). Production of β-galactosidase from recombinant Saccharomyces cerevisiae grown on lactose. *Journal of Chemical Technology & Biotechnology: International Research in Process, Environmental & Clean Technology, 79*(8), 809–815.

Domingues, L., Lima, N., & Teixeira, J. A. (2005). Aspergillus Niger β-galactosidase production by yeast in a continuous high cell density reactor. *Process Biochemistry, 40*(3–4), 1151–1154.

Duan, X., & Luan, S. (2023). Efficient secreted expression of natural intracellular β-galactosidase from bacillus aryabhattai via non-classical protein secretion pathway in Bacillus subtilis. *International Journal of Biological Macromolecules, 248*, 125758.

Duan, X., Hu, S., Qi, X., Gu, Z., & Wu, J. (2017). Optimal extracellular production of recombinant Bacillus circulans β-galactosidase in Escherichia coli BL21 (DE3). *Process Biochemistry, 53*, 17–24.

Eldin, M. M., Hassan, E. A., & Elaassar, M. R. (2005). β-Galactosidase covalent immobilization on the surface of alginate beads and its application in lactose hydrolysis. *Deutsch lebensmittel-Rundschau, 101*, 255–259.

Gupte, A. M., & Nair, J. S. (2010). β-Galactosidase production and ethanol fermentation from whey using Kluyveromyces marxianus NCIM 3551 Journal of Scientific & Industrial Research 69(11):855-859.

Haider, T., & Husain, Q. (2008). Concanavalin a layered calcium alginate–starch beads immobilized β galactosidase as a therapeutic agent for lactose intolerant patients. *International Journal of Pharmaceutics, 359*(1–2), 1–6.

Hsu, C. A., Yu, R. C., & Chou, C. C. (2005). Production of β-galactosidase by Bifidobacteria as influenced by various culture conditions. *International Journal of Food Microbiology, 104*(2), 197–206.

Hsu, C. A., Lee, S. L., & Chou, C. C. (2007). Enzymatic production of galactooligosaccharides by β-galactosidase from Bifidobacterium longum BCRC 15708. *Journal of Agricultural and Food Chemistry, 55*(6), 2225–2230.

Hu, X., Robin, S., O'Connell, S., Walsh, G., & Wall, J. G. (2010). Engineering of a fungal β-galactosidase to remove product inhibition by galactose. *Applied Microbiology and Biotechnology, 87*, 1773–1782.

Ishikawa, K., Kataoka, M., Yanamoto, T., Nakabayashi, M., Watanabe, M., Ishihara, S., & Yamaguchi, S. (2015). Crystal structure of β-galactosidase from Bacillus circulans ATCC 31382 (BgaD) and the construction of the thermophilic mutants. *The FEBS Journal, 282*(13), 2540–2552.

Jacobson, R. H., Zhang, X. J., DuBose, R. F., & Matthews, B. W. (1994). Three-dimensional structure of β-galactosidase from E. coli. *Nature, 369*(6483), 761–766.

Ji, D., Sims, I., Xu, M., Stewart, I., & Agyei, D. (2021). Production and identification of galacto-oligosaccharides from lactose using β-D-galactosidases from lactobacillus leichmannii 313. *Carbohydrate Polymer Technologies and Applications, 2*, 100038.

Jorgensen, F., Hansen, O., & Stougaard, P. (2001). High-efficiency synthesis of oligosaccharides with a truncated β-galactosidase from Bifidobacterium bifidum. *Applied Microbiology and Biotechnology, 57*, 647–652.

Juers, D. H., Jacobson, R. H., Wigley, D., Zhang, X. J., Huber, R. E., Tronrud, D. E., & Matthews, B. W. (2000). High resolution refinement of β-galactosidase in a new crystal form reveals multiple metal-binding sites and provides a structural basis for α-complementation. *Protein Science, 9*(9), 1685–1699.

Juers, D. H., Heightman, T. D., Vasella, A., McCarter, J. D., Mackenzie, L., Withers, S. G., & Matthews, B. W. (2001). A structural view of the action of *Escherichia coli* (lac Z) β-galactosidase. *Biochemistry, 40*(49), 14781–14794.

Jurado, E., Camacho, F., Luzon, G., & Vicaria, J. M. (2002). A new kinetic model proposed for enzymatic hydrolysis of lactose by a β-galactosidase from Kluyveromyces fragilis. *Enzyme and Microbial Technology, 31*(3), 300–309.

Kim, G. E., Lee, J. H., Jung, S. H., Seo, E. S., Jin, S. D., Kim, G. J., et al. (2010). Enzymatic synthesis and characterization of hydroquinone galactoside using Kluyveromyces lactis lactase. *Journal of Agricultural and Food Chemistry, 58*(17), 9492–9497.

Lu, L., & Xiao, M. (2017). Recent Progress on galactooligosaccharides synthesis by microbial β-galactosidase. In *Functional carbohydrates: Development, characterization, and biomanufacture* (pp. 147–182).

Lu, L., Xu, L., Guo, Y., Zhang, D., Qi, T., Jin, L., et al. (2015a). Glycosylation of phenolic compounds by the site-mutated β-galactosidase from lactobacillus bulgaricus L3. *PLoS One, 10*(3), e0121445.

Lu, L., Guo, Y., Xu, L., Qi, T., Jin, L., Xu, L., & Xiao, M. (2015b). Galactosylation of caffeic acid by an engineered β-galactosidase. *Drug Discoveries & Therapeutics, 9*(2), 123–128.

Lu, L., Guo, L., Wang, K., Liu, Y., & Xiao, M. (2020). β-Galactosidases: A great tool for synthesizing galactose-containing carbohydrates. *Biotechnology Advances, 39*, 107465.

Marín-Navarro, J., Talens-Perales, D., Oude-Vrielink, A., Cañada, F. J., & Polaina, J. (2014). Immobilization of thermostable β-galactosidase on epoxy support and its use for lactose hydrolysis and galactooligosaccharides biosynthesis. *World Journal of Microbiology and Biotechnology, 30*, 989–998.

Martinez, A., Ramirez, O. T., & Valle, F. (1997). Improvement of culture conditions to overproduce β-galactosidase from Escherichia coli in Bacillus subtilis. *Applied Microbiology and Biotechnology, 47*, 40–45.

Matioli, G., De Moraes, F. F., & Zanin, G. M. (2003). Operational stability and kinetics of lactose hydrolysis by β-galactosidase from Kluyveromyces fragilis. *Acta Scientiarum, 25*(1), 7–12.
Matthews, B. W. (2005). The structure of E. Coli β-galactosidase. *Comptes Rendus Biologies, 328*(6), 549–556.
Movahedpour, A., Ahmadi, N., Ghalamfarsa, F., Ghesmati, Z., Khalifeh, M., Maleksabet, A., et al. (2022). β-Galactosidase: From its source and applications to its recombinant form. *Biotechnology and Applied Biochemistry, 69*(2), 612–628.
Nguyen, T. T., Nguyen, H. M., Geiger, B., Mathiesen, G., Eijsink, V. G., Peterbauer, C. K., et al. (2015). Heterologous expression of a recombinant lactobacillal β-galactosidase in lactobacillus plantarum: Effect of different parameters on the sakacin P-based expression system. *Microbial Cell Factories, 14*(1), 1–11.
Nguyen, T. H., Splechtna, B., Yamabhai, M., Haltrich, D., & Peterbauer, C. (2007). Cloning and expression of the β-galactosidase genes from Lactobacillus reuteri in Escherichia coli. *Journal of Biotechnology, 129*(4), 581–591.
Nijland, R., & Kuipers, O. P. (2008). Optimization of protein secretion by Bacillus subtilis. *Recent Patents on Biotechnology, 2*(2), 79–87.
Oliveira, C., Teixeira, J. A., Lima, N., Da Silva, N. A., & Domingues, L. (2007). Development of stable flocculent Saccharomyces cerevisiae strain for continuous aspergillus Niger β-galactosidase production. *Journal of Bioscience and Bioengineering, 103*(4), 318–324.
Oliveira, C., Guimaraes, P. M., & Domingues, L. (2011). Recombinant microbial systems for improved β-galactosidase production and biotechnological applications. *Biotechnology Advances, 29*(6), 600–609.
Panesar, P. S., Panesar, R., Singh, R. S., Kennedy, J. F., & Kumar, H. (2006). Microbial production, immobilization and applications of β-D-galactosidase. *Journal of Chemical Technology & Biotechnology: International Research in Process, Environmental & Clean Technology, 81*(4), 530–543.
Panesar, P. S., Kumari, S., & Panesar, R. (2010). Potential applications of immobilized β-galactosidase in food processing industries. *Enzyme Research, 2010*, 473137.
Pessela, B. C., Vian, A., Mateo, C., Fernandez-Lafuente, R., Garcia, J. L., Guisan, J. M., & Carrascosa, A. V. (2003). Overproduction of Thermus sp. strain T2 β-galactosidase in Escherichia coli and preparation by using tailor-made metal chelate supports. *Applied and Environmental Microbiology, 69*(4), 1967–1972.
Placier, G., Watzlawick, H., Rabiller, C., & Mattes, R. (2009). Evolved β-galactosidases from Geobacillus stearothermophilus with improved transgalactosylation yield for galacto-oligosaccharide production. *Applied and Environmental Microbiology, 75*(19), 6312–6321.
Raguin, A., Stansfield, I., & Romano, M. C. (2023). ExpressInHost: A codon tuning tool for the expression of recombinant proteins in host microorganisms. *Journal of Open Research Software*.
Rico-Díaz, A., Álvarez-Cao, M. E., Escuder-Rodríguez, J. J., González-Siso, M. I., Cerdán, M. E., & Becerra, M. (2017). Rational mutagenesis by engineering disulphide bonds improves Kluyveromyces lactis beta-galactosidase for high-temperature industrial applications. *Scientific Reports, 7*(1), 45535.
Roy, I., & Gupta, M. N. (2003). Lactose hydrolysis by Lactozym™ immobilized on cellulose beads in batch and fluidized bed modes. *Process Biochemistry, 39*(3), 325–332.
Selvarajan, E., & Mohanasrinivasan, V. (2015). Kinetic studies on exploring lactose hydrolysis potential of β galactosidase extracted from lactobacillus plantarum HF571129. *Journal of Food Science and Technology, 52*, 6206–6217.
Simonen, M., & Palva, I. (1993). Protein secretion in bacillus species. *Microbiological Reviews, 57*(1), 109–137.
Urrutia, P., Bernal, C., Wilson, L., & Illanes, A. (2014). Improvement of chitosan derivatization for the immobilization of bacillus circulans β-galactosidase and its further application in galacto-oligosaccharide synthesis. *Journal of Agricultural and Food Chemistry, 62*(41), 10126–10135.
Valuates Reports (2024). Global beta-galactosidase market research report.

Wallenfels, K., & Weil, R. (1972). 20 β-Galactosidase. In *The enzymes* (Vol. 7, pp. 617–663). Academic.

Welsch, N., Homuth, G., & Schweder, T. (2012). Suitability of different β-galactosidases as reporter enzymes in Bacillus subtilis. *Applied Microbiology and Biotechnology, 93*, 381–392.

Westers, L., Westers, H., & Quax, W. J. (2004). Bacillus subtilis as cell factory for pharmaceutical proteins: A biotechnological approach to optimize the host organism. Biochimica et Biophysica Acta (BBA)-molecular. *Cell Research, 1694*(1–3), 299–310.

Wu, Y., Yuan, S., Chen, S., Wu, D., Chen, J., & Wu, J. (2013). Enhancing the production of galacto-oligosaccharides by mutagenesis of Sulfolobus solfataricus β-galactosidase. *Food Chemistry, 138*(2–3), 1588–1595.

Xiong, A. S., Peng, R. H., Zhuang, J., Li, X., Xue, Y., Liu, J. G., et al. (2007). Directed evolution of a beta-galactosidase from Pyrococcus woesei resulting in increased thermostable beta-glucuronidase activity. *Applied Microbiology and Biotechnology, 77*, 569–578.

Yu, L., & O'sullivan, D. J. (2014). Production of galactooligosaccharides using a hyperthermophilic β-galactosidase in permeabilized whole cells of Lactococcus lactis. *Journal of Dairy Science, 97*(2), 694–703.

Zhang, J. H., Dawes, G., & Stemmer, W. P. (1997). Directed evolution of a fucosidase from a galactosidase by DNA shuffling and screening. *Proceedings of the National Academy of Sciences, 94*(9), 4504–4509.

Zhang, Z., Zhang, F., Song, L., Sun, N., Guan, W., Liu, B., et al. (2018). Site-directed mutation of β-galactosidase from Aspergillus candidus to reduce galactose inhibition in lactose hydrolysis. *3 Biotech, 8*, 1–7.

Zhou, Q. Z., & Chen, X. D. (2001). Effects of temperature and pH on the catalytic activity of the immobilized β-galactosidase from Kluyveromyces lactis. *Biochemical Engineering Journal, 9*(1), 33–40.

Chapter 4
Approaches for Lactose Hydrolysis

M. H. Sathish Kumar, B. Ashritha, and C. T. Manoj Kumar

4.1 Introduction

Lactose is the primary constituent of whey, which is likely to be the least valued component and presents the greatest challenge in terms of utilization. Lactose constitutes for approximately 70–72% of the total solids in both sweet and acid whey. During the stage of infancy, it serves as an essential and crucial source of energy, supplying over 50% of the overall energy requirements of infants (Macwan et al., 2016) Additionally, it plays a crucial role in biological processes such as promoting the development of bifidobacteria in the gastrointestinal tract and providing galactose, a vital component for the synthesis of cerebral galactolipids (Lv & Yu, 2014). Lactose, along with other sugars, functions as a transporter for minerals by forming ligands, thereby facilitating their absorption. Lactose, along with other sugars, serves as a transporter for minerals by forming ligands, which assists in their absorption (Schaafsma, 2008).

In order for lactose to be assimilated, it must undergo hydrolysis, resulting in the formation of its individual components—glucose and galactose. The primary catalyst for this process is the β-galactosidase enzyme (β-D-galactoside galactohydrolase, E.C. 3.2.1.23), often known as lactase. The enzyme is attached to the brush border membrane of the mucosa in the small intestine. It exhibits its highest level of activity in the jejunum and a lower level in the duodenum and ileum. The monosaccharides are efficiently absorbed by the small intestine upon the hydrolysis of lactose. Subsequently, they are delivered to the liver through the portal vein and ultimately employed as a source of energy or as building blocks in the body. Following absorption, galactose undergoes conversion into glucose within the liver via the Leloir route, which encompasses the activity of three enzymes

M. H. Sathish Kumar (✉) · B. Ashritha · C. T. Manoj Kumar
Dairy Technology Section, Southern Regional Station, ICAR-National Dairy Research Institute, Bengaluru, Karnataka, India

P. S. Rao et al. (eds.), *Lactose Hydrolysis in Dairy Products*,
https://doi.org/10.1007/978-3-031-78207-7_4

(galactokinase, galactose-1-P uridylyltransferase, and UDP-galactose 4-epimerase). Lack of any of these enzymes can result in metabolic disruptions referred to as lactose intolerance and galactosemia (Perotti, 2012). The minimal concentration of lactose solids in milk, whey, and permeate (about 5%), combined with its cost disadvantage as a raw material for numerous fermentation processes, and the relatively cheap price for pure lactose, frequently hinders the economic viability of commercially utilizing lactose-containing by-products. Conversely, the elevated biological oxygen demand of lactose discharges poses challenges and incurs significant costs for disposal. In this regard, Lactose hydrolysis presents the dairy sector with the chance to solve these challenges by enabling the creation of a variety of products that possess distinct desirable functional characteristics and marketing benefits. During hydrolysis, the disaccharide lactose undergoes a chemical or enzymatic conversion into its two constituent monosaccharides, glucose and galactose (Oliveira et al., 2023). This method can be utilized for all dairy fluids that include lactose, such as milk, whey, and permeates obtained from the ultrafiltration of milk or whey. Nevertheless, certain hydrolysis systems are impractical for processing all of these streams. It is crucial for potential users of this technology to be fully informed about its limitations. Lactose hydrolysis can be achieved through acid hydrolysis, which involves heating at low pH, or enzymatic catalysis using the enzymes lactase or β-D-galactosidase. The enzyme can be either free in solution or immobilized to solid support to sustain continuous production process (Hartmeier et al., 2012). Multiple approaches to lactose hydrolysis is given as Fig. 4.1.

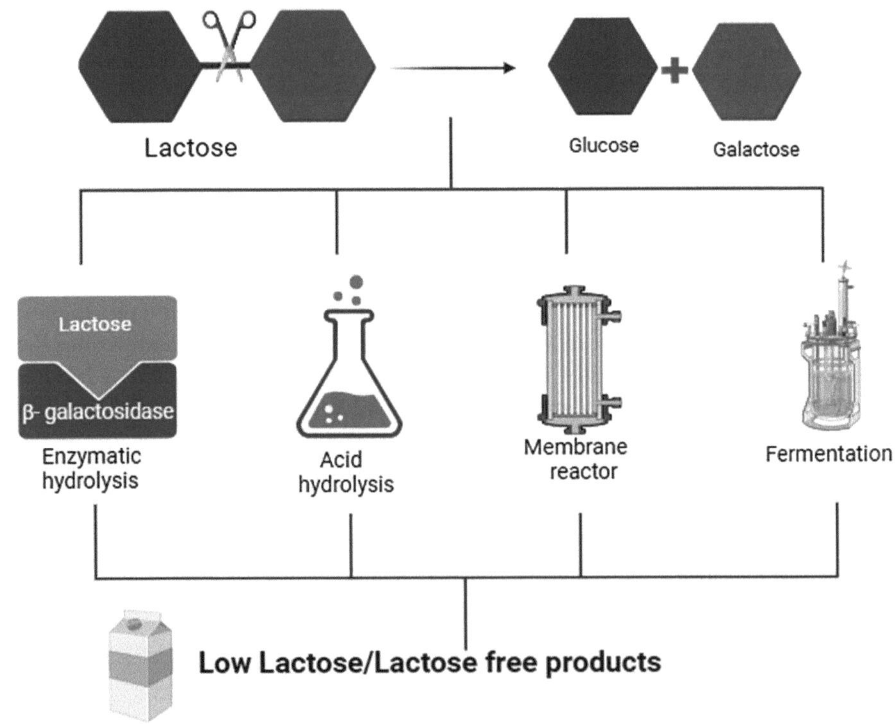

Fig. 4.1 Multiple approach to perform lactose hydrolysis

4.2 Acid Hydrolysis

Lactose hydrolysis by acid has been demonstrated by Ramsdell and Webb (1945); Ney and Wirotama (1970); Coughlin and Nickerson (1975). Acid can only be used for lactose hydrolysis in streams that are free of protein, such as ultrafiltration (UF) permeate of milk or whey. The pH adjustment can be achieved either by directly adding acid or by treating the permeate with an ion exchange resin. Usually, the pH is reduced to a value of 1.2 and the temperature is set to 150 °C for a brief duration. Hydrochloride has traditionally been the acid of choice for the reaction, while H_2SO_4 has the benefit of allowing the removal of the anion through precipitation with salt. This will decrease the level of salts and enhance the quality of flavor. This acid could be more economical. The hydrolyzed product has a brown color and requires neutralization, demineralization, and decolorization prior to utilization. The corrosive nature of the circumstances poses a significant engineering design obstacle in this process. Therefore, it has not been widely embraced for commercial use (Zadow, 1986). Industrially, this method has exclusively been employed for solutions containing solely pure lactose. The process of acid hydrolysis of lactose primarily produces glucose and galactose as the major products, with reaction yields ranging from 90% to 99% (Cheng & Martínez-Monteagudo, 2019). The process of oligosaccharide production resulting from the hydrolysis of lactose using mineral acid or a highly acidic cation exchange resin was documented by Aronson (1952); Huh et al. (1991) and several others. Reports have indicated the formation of disaccharides with diverse forms, including α and β configurations. Similar events were noted in lactose during the process of acidic hydrolysis. Isomaltose and gentiobiose are disaccharides that are derived from glucose, while 6-O-α-D-galactopyranosyl--D-galactose and 6-O-β-D-galactopyranosyl-D-galactose are derived from galactose (Stanek et al., 1965).

4.3 Enzymatic Hydrolysis

Lactose hydrolysis is a precise technique that causes minimal damage to the nutritious elements in milk. Enzymatic hydrolysis is more favorable than acidic hydrolysis due to its ability to operate in milder pH and temperature circumstances, without resulting in undesirable odors, aromas, and colors. In addition, the use of an acidic technique might lead to the denaturation of proteins that may be present in the lactose solution, resulting in the production of unwanted by-products (Demirhan et al., 2010) Nevertheless, in the industrial sector, exclusively enzymes derived from microbial sources are employed for the process of lactose hydrolysis (Husain, 2010). This is a highly precise biotechnological procedure that utilizes either a single enzyme, numerous enzymes, or immobilized β-galactosidases. β-galactosidases are a widely distributed group of enzymes called glycoside hydrolases, which can break down the glycosidic bond in lactose to produce glucose and galactose. These are present in microorganisms such as bacteria, fungi, and yeasts, as well as in plants, particularly in almonds, peaches, apricots, and apples. They can also be found in animal organs (Haider & Husain, 2007). Many organisms naturally produce these substances, but industries commonly use enzymes obtained from microbes due to their higher production yield and rapid multiplication rate. Microbial enzymes are cost-effective, require minimal space, have enhanced substrate versatility, better stability, are regularly available, easy to handle, and can be produced on a large scale compared to enzymes from plant and animal sources (Kalathinathan et al., 2023). Typically, enzymes utilized in the food and pharmaceutical sectors are obtained from microorganisms that are considered safe for human consumption, referred to as GRAS (Generally Recognized as Safe). Fungi such as *Aspergillus flavus*, *Aspergillus foetidus*, *Aspergillus niger*, *Aspergillus phoenicis*, *Aspergillus oryzae*, *Mucor pusillus*, *Mucor miehei*, and *Neurospora crassa* are examples of fungal sources that produce β-galactosidase. Typically, enzymes that are successful in breaking down lactose in acidic products like whey have a pH optimum between 2.5 and 5.4. The primary industrial enzymes are derived from mold Aspergillus sp. and yeast Kluyveromyces sp. β-galactosidase derived from *Kluyveromyces lactis* is extensively employed in various applications (Singh & Sambyal, 2023). The characteristics of β-galactosidase, including specificity, structure, molecular weight, amino acid sequence, active site position, and optimal pH and temperature, vary greatly depending on the microbial source. This enzyme typically resides within the cell in bacteria and yeasts, but in fungi it can be found either within or outside the cell. The temperature and pH at which lactases function best are the main criteria that determine the range of potential applications. Typically, lactases derived from Kluyveromyces, which have a pH range of 6.5–7.5 for optimal activity, are employed to break down lactose in milk at a moderate temperature of 30–40 °C due to their sensitivity to heat (Perotti, 2012). These working circumstances can promote the proliferation of microorganisms. Fungal β-galactosidases that operate optimally at acidic pH levels have the added benefit of exhibiting greater thermal stability compared to those derived from yeasts. Consequently, they can be effectively employed at temperatures as high as 50 °C. Elevated temperatures and an acidic atmosphere effectively inhibit microbial development (Abd-El-Salam et al., 2020). Divalent cations, such as Mn^{2+} and Mg^{2+}, have the potential to enhance the activity of enzymes. On the other hand, the impact of monovalent cations, such Na^+ and K^+, can vary depending on the enzyme's source (Dekker et al., 2019).

4.3.1 Batch Process

The first technique is the batch process, which involves adding neutral lactase to raw milk and then agitating it gently for almost 24 h to prevent creaming. Additionally, this procedure is carried out in a temperature range of 4–8 °C in order to prevent the growth of microorganisms, as the milk has not yet been sterilized. Following the incubation process, milk undergoes pasteurization, homogenization, and packaging

(Dekker et al., 2019). In the second method, a sterile enzyme is added after heat treatment of milk and just before packaging (Troise et al., 2016). The finished product does not include any remaining enzyme activity as the enzyme is deactivated throughout the sterilization/pasteurization process. The enzyme dosage must be adequate to achieve the desired threshold for lactose-free within the specified incubation period and at a low temperature. Consequently, the amount of enzyme used is comparatively elevated. The enzymes chosen for this process are specifically designed for their elevated efficiency at a neutral pH and low temperature.

The process control is robust as it allows for the measurement of lactose conversion in the tank and enables adjustments to be made to the enzyme dosage or incubation time as required. The batch incubation necessitates the utilization of a tank within the industrial premises and the retention of the milk for a duration of 1 day. Hence, the process might present challenges for certain manufacturers, particularly those with high levels of production. A lactase enzyme with enhanced specific activity under these specific conditions, such as Maxilact® Smart, has the potential to reduce production time and thereby improve the factory's overall throughput. In order to mitigate the usual increase of sweetness that occurs following lactose hydrolysis and restore a more standard palatability, procedures such as ultrafiltration, nanofiltration, or chromatography (in conjunction with the hydrolysis of the remaining lactose) are employed (Jelen & Tossavainen, 2003); Harju et al., 2012).

4.3.1.1 Single Use Enzyme

Unlike continuous processes where enzymes are recycled, in the single-use approach, the enzyme is utilized only once. After completing the hydrolysis reaction, the enzyme is discarded. The method that employs the soluble or free enzyme and is the most widely applied because of its feasibility of implementation at an industrial scale. When utilizing a single enzyme for the process of lactose hydrolysis, it is important to take into account various elements such as substrate concentration, operational pH, maximum temperature, contact time limit, enzyme activity, and cost. The most economical approach is to employ small quantities of soluble enzyme and provide a prolonged duration of exposure, often lasting 24–48 h. Refrigerating the substrate is essential to preserve it for an extended duration, hence decreasing the enzyme's activity. This technique necessitates large storage tanks and may lead to microbial issues with certain substrates.

Alternatively, the duration of hydrolysis can be greatly decreased by increasing the amount of enzyme and by raising the temperature to a range closer to the optimal temperature for the enzyme, around 40–50 °C. Under these conditions, the holding durations typically ranges from 3 to 5 h, contingent upon the quantity and effectiveness of the enzyme introduced (Dekker et al., 2019). It is important to closely observe the growth of microorganisms in the substrate during the holding period. Given that the enzyme's activity is positively correlated with temperature, it is unlikely that prolonging the holding times under refrigerated settings will yield significant economic advantages. Researchers have found the best conditions for using these enzymes to get a high level of lactose hydrolysis. So, in order to get a high level of lactose hydrolysis (more than 80%), the following lactose hydrolysis plans were made: Cold lactose hydrolysis should be conducted at 4–6 °C with 0.3% enzyme from *B. licheniformis* for 4 h or with 0.3% enzyme from *K. lactis* for 12 h. Lactose hydrolysis at 38–40 °C with 0.3% *B. licheniformis produced enzyme* results in over 90% hydrolysis within 2 h. The ideal temperature range for *K. lactis* enzyme is 38–40 °C, with a concentration of 0.15% and a hydrolysis period of 3–4 h. The leftover lactase enzyme is inactivated at 70 °C during pasteurization (Popescu et al., 2021).

It is crucial to note that if the hydrolysis time is extended due to reduced activity at low temperatures, it will be necessary to employ additional reactors to achieve the target output. The process of lactose hydrolysis typically does not result in the complete transformation of monosaccharides. This is mostly owing to the inhibitory effect of galactose on the reaction, as well as the occurrence of simultaneous side reactions that generate oligosaccharides (Gosling et al., 2010). Oligosaccharides are synthesized through transglycosylation, a step facilitated by the transferase activity of enzymes. The enzyme-galactose intermediate has a reaction with a different sugar molecule (such as lactose, glucose, or galactose). Initially, oligosaccharides, primarily made up of galacto-oligosaccharides (GOS), were seen as unwanted by-products of lactose hydrolysis. However, recent studies have indicated that these compounds are not easily broken down by digestive enzymes and actually have a positive impact on the health of the human intestines. They promote the growth of bifidobacteria, which is beneficial for intestinal health (known as bifidogenic factor) (Park & Oh, 2010). Evaluation of β-galactosidase activity for lactose hydrolysis (DH) and galactooligosaccharide (GOS) formation at 7 °C has been carried out by Singh et al. (2021). Researchers found that low lactose concentrated skim milk with GOS can be manufactured sing 9.53 U/mL of β-galactosidase at 7 °C which resulted in 97.06 DH (%) and 9.45% GOS after 12 h. β-galactosidase from *K. lactis* was found to be feasible for both lactose hydrolysis as well as GOS formation in concentrated skim milk.

4.3.1.2 Multi-enzyme Approach

β-Galactosidase is a hydrolase that cleaves the o-glucosyl group of lactose. The general mechanism of enzymatic lactose hydrolysis has a transgalactosylic nature, involving a multitude of sequential reactions with disaccharides (other than lactose) and higher saccharides, collectively named galacto-oligosaccharides (GOS), as intermediate products (Wallenfels & Malhotra, 1962). It is well known that oligosaccharides can be formed from monosaccharides by the action of mineral acids (chemical synthesis). This process, known as "reversion," explains the production of oligosaccharides during acidic hydrolysis of lactose, first observed in the 1950s (Aronson, 1952). The conditions suitable for oligosaccharide production during acidic hydrolysis of lactose and the resulting oligosaccharide structures formed have been well studied. It was reported that there is formation of a complex mixture of disaccharides and trisaccharides, with a variety of linkages with α- and β-anomeric configurations, and anhydro-sugars, as a result of this chemical process (Huh et al., 1991).

The limited utilization of this GOS production technique on a broad scale can be attributed to the absence of product specificity and the rigorous conditions employed during acidic hydrolysis of lactose. The enzyme's ability to create and break down a range of oligosaccharides through transferase activity has garnered interest from researchers. This is due to the beneficial impact of oligosaccharides on the proliferation of desirable microorganisms in the intestines. Furthermore, the transferase reaction can be utilized to covalently bond galactose to various chemical compounds, hence offering prospective applications in the synthesis of food components, medicines, and other biologically active molecules. Cultures with significant transgalactosylation activity have been identified across many cultures. The synthesis of novel oligosaccharide combinations was achieved through the utilization of β-D-galactosidases derived from probiotic bacteria, as described by Rabiu et al. (2001). The bacteria were employed in glycosyl transfer processes for the production of oligosaccharides from lactose, resulting in oligosaccharide yields ranging from 24.7 to 47.6%. The β-D-galactosidase enzyme derived from *Bifidobacterium bifidum* shown efficacy in transgalactosylation processes, enabling the synthesis of oligosaccharides even when lactose contents were low (Jørgensen et al., 2001). In another study, Rico-Rodríguez et al. (2020) introduced a multi-enzyme system consisting of glucose oxidase, β-galactosidase, and catalase into cheese whey. They then subjected the reaction to ultrasound at an amplitude of 30%. It was best to use lactose solution in multienzymatic systems with ultrasound (30% amplitude) and get the highest rates of GOS production with the lowest glucose concentration. This was also the case when glucose oxidase was added after β-galactosidase treatment for 1 h. Fischer and Kleinschmidt (2018) conducted research on the synthesis of GOS (galactooligosaccharides) utilizing a multi-enzyme system consisting of glucose oxidase, β-galactosidase, and catalase. They observed that glucose was simultaneously eliminated during the process of transgalactosylation.

4.3.2 Aseptic Process (Post-hydrolysis)

During the aseptic process, the milk undergoes sterilization using the UHT procedure. Subsequently, a sterile lactase preparation is introduced into the milk shortly before it is packaged (Dahlqvist et al., 1977). The lactose conversion in the milk will occur within the milk container. Due to the typical practice of subjecting UHT milk to an incubation period of about 3 days at room temperature, there is ample opportunity for full hydrolysis to occur before the milk is sent to the store. Considering the lack of such incubation period for pasteurized milk, the aseptic procedure is unnecessary for the production of this particular lactose-free milk variant. There are essentially two distinct methods for acquiring a sterile lactase.

(a) In the first method, the lactase enzyme undergoes presterilization by the enzyme producer. Additionally, the sterile injection necessitates the use of certain sterile dosage equipment, such as the Tetrapak (Lund, Sweden) Flexdos® system.
(b) In the second step, the enzyme that is not sterile is passed through a filter to remove any contaminants right before it is added to the sterile milk at the dairy factory. This can be done using a system like the Tetrapak Aldose®.

Due to the increased incubation time and temperature, the dosage of the enzyme can be significantly reduced in comparison to the batch procedure. Process control is lacking as the enzyme is solely active in the packaged milk. For instance, the temperature in warehouses without temperature control may vary between summer and winter. The dairy producer should consider these factors when determining the dosage of the enzyme. The aseptic procedure necessitates specialized equipment and incurs consumable cost. Particularly for in-factory screening, it demands highly proficient operators to prevent microbial contamination of the milk during lactase injection. Nevertheless, when effectively arranged, the process can be executed in a fully continuous manner, which is a significant benefit for companies with a need for a large volume of output (Dekker et al., 2019).

4.3.3 Immobilized Enzyme System

A more contemporary method that shows promise in lactose hydrolysis involves employing cell cultures that produce β-galactosidases directly, eliminating the need for additional purification. The usage of soluble or free enzymes is the most commonly utilized technology due to its practicality for large-scale industrial deployment. Purified β-galactosidase is introduced into the solution and subsequently inactivated after a single use, which is a drawback. The immobilization of enzymes involves physically confining the enzyme to a solid matrix while preserving its ability to catalyze reactions. This method has been developed to enhance reaction control and to enable reuse of enzyme. The soluble enzyme is often employed in batch processes, whereas the immobilized form offers the added benefit of being usable in both batch-wise and continuous operations (Panesar et al., 2010). In the 1970s, Snam Progetti, an Italian company, created an immobilized galactosidase enzyme derived from *Saccharomyces lactis*. This enzyme was trapped between cellulose triacetate fibers and proved to be effective in the hydrolysis operations of Centrale del Latte in Italy and Snow Brand Milk Products in Japan. Immobilization has been demonstrated to enhance the stability of β-galactosidase, facilitate reusability, and decrease processing time in the food applications. β-galactosidase immobilized in a fixed state may constantly break down lactose in dairy products, resulting in a substantial enhancement of lactose hydrlysis efficiency (Arijit Nath et al., 2013). There are two main reasons why microorganisms are able to survive in milk. Firstly, milk has a neutral pH, which is favourable for their growth. Secondly, the proteins in milk easily attach to the surface of immobilized β-galactosidase. Furthermore, the selection of a technique for enzyme immobilization is crucial since it directly impacts the enzyme's performance due to the interactions between the carrier and the enzyme. β-D-galactosidase has been immobilized using diverse techniques including physical adsorption, gel entrapment, covalent attachment onto various substrates, or a combination of these methods (Fig. 4.1). Various immobilized enzyme systems have been studied for the purpose of breaking down lactose in milk and whey by hydrolysis (Deng et al., 2024).

4.3.3.1 Physical Adsorption

Physical adsorption is a way of fixing an enzyme onto a carrier that is not soluble in water. The enzyme is held on the carrier's surface by physical forces, namely van der Waals forces. This is considered the simplest kind of immobilization. Often, there are other factors at play to moderate connection between the carrier and biocatalyst, specifically hydrophobic interactions, hydrogen bonding, and heteropolar (ionic) bonds (Hartmeier et al., 2012). Common materials used for immobilizing β-galactosidase include alginate, chitosan, silica, and resin, among others. Emerging materials are consistently advancing, with their particle size typically falling within the nanoscale range. This characteristic yields a significantly higher surface-to-volume ratio, resulting in enhanced surface reactivity and enhanced catalytic efficacy. The primary nanomaterials utilized for immobilizing β-galactosidase are carbon nanotubes, silicon dioxide nanoparticles, nanodiamonds, silver nanoparticles, and zinc oxide nanoparticles (Ureta et al., 2021). β-galactosidase has been attached to porous cellulose acetate fibers by means of adsorption. At first immobilized β-galactosidase enzyme capable of hydrolyzing lactose in milk for industrial production of low-lactose milk. By implementing further optimizations, the residual amount of enzyme in the final product was successfully reduced (Harju, 2004). The immobilization of β-galactosidase on various carriers has been achieved through effective physical adsorption. This method offers several benefits, including an easy process, gentle conditions, a broad selection of potential carriers, and the ability to replicate results. However, a drawback of this method is that the β-galactosidase enzyme, once immobilized, is susceptible to environmental influences during usage, such as fluctuations in temperature, pH, and ionic strength. These factors can lead to the detachment of the β-galactosidase enzyme. Immobilized enzyme exhibits more activity than free one, mostly due to its increased affinity for the substrate. According to a recent study conducted by Shafi et al. (2021) immobilized β-galactosidase shown a retention of 85% of its activity after undergoing 10 repeated applications. Additionally, it exhibited the ability to decrease the lactose level in milk by 89%.

4.3.3.2 Covalent Binding

Covalent binding refers to the process of enzymes being securely attached to a support surface by the production of covalent bonds. Enzyme molecules attach to support material by specific functional groups, including amino, carboxyl, hydroxyl, and sulfhydryl groups. The immobilization of the enzyme through covalent bonding relies on modifying the surface of the support material using activation processes. In order to create a highly effective interaction between enzymes and their support, the functional groups of the support material undergo modifications that result in the formation of reactive intermediates (Souza et al., 2018). The functional groups on the support material are often activated through the use of chemical reagents such as cyanogen bromide, carbodimide, and glutaraldehyde (Panesar et al., 2010). Optimizing immobilization parameters, such as reaction duration, pH, temperature, and inhibitor concentration, is crucial for maximizing the formation of covalent bonds between the enzyme and carrier (Ureta et al., 2021). β-galactosidase has been covalently attached to inorganic supports, including graphite, fiber substances, oxides, magnetic materials, and glass microspheres (Deng et al., 2024). Natural biopolymer carriers are considered to be safer for usage in the food sector when compared to inorganic carriers (Elnashar & Yassin, 2009). Chitosan, a naturally occurring substance, is frequently employed as a carrier for the immobilization of β-galactosidase. This has been documented in various studies, including those by Malecki et al. (2013); (Klein et al., 2016); Urrutia et al. (2018), and Katrolia et al. (2019).

4.3.3.3 Cross-Linking

The crosslinking approach for enzyme immobilization involves the employment of a crosslinking chemical to establish covalent connections between enzymes. The crosslinking agent possesses two reactive ends that adhere to amino acids within the enzyme. Carrier-free immobilization refers to the process of cross-linking enzyme molecules without the use of carriers. This method allows for the affordable immobilization of enzymes while maintaining a high specific activity (Duan et al., 2022). The enzyme can undergo self-cross-linking to create a three-dimensional network when exposed to crosslinking chemicals (Nath et al., 2014). In this scenario, the use of a carrier is unnecessary, therefore eliminating the issue of diffusion and mass transfer constraints. Nevertheless, the self-crosslinking process may lead to the formation of enzyme clusters, which subsequently leads to a decrease in enzymatic activity (Damin et al., 2021). Enzymes and carriers can form covalent bonds through crosslinking using bi- or multifunctional crosslinkers such as glutaraldehyde, hexamethylene diisocyanate, diisocyanate, glyoxal, toluene diisocyanate, and genipin. Immobilized derivatives are produced by crosslinking a crosslinking agent with an enzymatic preparation. The enzymatic preparation can be in the form of solubilized, crystallized, atomized, or aggregated enzymes, resulting in the formation of cross-linked enzyme, cross-linked enzyme crystals, cross-linked spray-dried enzyme, or cross-linked enzyme aggregates, respectively. Crosslinking chemicals serve the dual purpose of activating the carrier to form a bond with the enzyme and also creating connections inside or between enzyme molecules, thereby enhancing the stability of the immobilized enzyme. The immobilized β-galactosidase shown superior efficiency in lactose hydrolysis in milk, as compared to free one (Ansari & Husain, 2012). A study conducted by Haider and Husain (2007) examined the performance of immobilized β-galactosidase in both crosslinked and non-cross-linked systems. The lactose conversion rate was found to be 77% in the crosslinked system and 54% in the non-cross-linked system. Additionally, the β-galactosidase activity retention rate was 86% in the crosslinked system vis-a-vis 30% in the non-cross-linked system after a storage period of 2 months.

4.3.4 Encapsulated Enzyme System

Microencapsulation is a technique employed by the food and pharmaceutical sectors to enclose solid, liquid, or gaseous substances. An advantage of this technology is its ability to release active substances at precise rates in certain environments, while also safeguarding them from interactions with other compounds in food and preventing oxidation. Microcapsules enhance the durability of delicate materials during processing and packaging, while also enhancing the taste, odor, stability, nutritional content, and visual appeal of the products (Ugidos-Rodríguez et al., 2018). Enzyme immobilization via entrapment offers many benefits, including cost-effectiveness, simplicity, and suitability for a wide range of enzymes. Additionally, this method employs renewable biopolymers such as alginate, chitosan, and cellulose as matrices. These biopolymers are not only abundant but also nontoxic, biocompatible, and biodegradable. The process of microencapsulating lactase has been developed as a solution to address the increasing need for natural alternatives. In contrast to synthetic polymers, natural polymers including alginate, carrageenan, pectin, chitosan, and gelatin are better suited for creating hydrogels to immobilize enzymes. This is because they are non-toxic and readily available in the food sector (Estevinho et al., 2014). The encapsulation techniques encompass several methods such as spray drying, extrusion, emulsion, spray chilling (or spray cooling), coacervation, cocrystallization, among others (Waghmare, 2020). The enzyme β-galactosidase was encapsulated in hydroxypropyl methylcellulose phthalate and added to milk at a concentration of roughly 1% (wt/vol). This encapsulation process allowed for controlled release of the enzyme, protected it against hydrolysis, and resulted in a retention rate of 81.18% after 12 days of refrigeration (Ahn et al., 2019).

Alginate is orally non-toxic and has a protective impact on the mucosal membrane of the upper gastrointestinal tract. Dried alginate beads have the ability to undergo re-swelling, which enables them to function as controlled-release devices (Bhardwaj et al., 2000). Panesar et al. (2006) investigated the manufacturing of lactose-hydrolyzed milk using *Kluyveromyces marxianus* NCIM 3465 cells that were permeabilized and served as a source of β-Dgalactosidase. The highest level of lactose hydrolysis (87.9%) was achieved by immobilizing yeast cells in alginate with a concentration of 2% (w/v), using beads with a size of 2.90 mm. This was done at a temperature of 30 °C, with agitation at 80 rpm, and after 150 min of incubation. A separate investigation involved the combination of alginate and gelatin to immobilize β-galactosidase. The immobilized β-galactosidase exhibited remarkable stability, with no decline in activity observed over a period of 35 days. Additionally, the immobilized β-galactosidase shown wider ranges of optimum temperature and pH compared to the free one (Tanriseven & Doğan, 2002). Zhang et al. (2016) conducted a study to examine the capability of carrageenan hydrogel beads in encapsulating lactase. The hydrogel beads were produced by infusing an aqueous solution containing β-galactosidase (26 U) and carrageenan (1%) into a hardening solution (5% potassium chloride). Approximately 63% of the lactase was initially enclosed within the hydrogel beads. The activity of encapsulated lactase was greater than that of the unbound enzyme under various pH and temperature settings. This can be due to the stability of the enzyme structure by K+ ions present in the carrageenan beads. The study conducted by Cargnin et al. (2021) demonstrated that pectin-based hydrogels have the potential to serve as solid supports for enzyme encapsulation and lactose hydrolysis. The catalytic activity of the encapsulated lactase was intact even after three consecutive cycles of lactose hydrolysis in UHT-milk. The utilization of pectin-alginate hydrogel for the encapsulation of β-galactosidase from *A. oryzae* enhanced the immobilization characteristics of the enzyme as well as the operating parameters (Paiva et al., 2023). The encapsulated enzyme demonstrated the potential for reusability in 8 cycles for the hydrolysis of lactose in milk whey and 7 cycles for the hydrolysis of lactose in whole milk. This has the potential to significantly decrease the operational expenses of milk processing companies. As extreme temperatures during spray drying (SD) have adverse effects on bioactive compounds. Freeze-drying, often known as FD, is a method that can effectively remove moisture from delicate molecules by subjecting them to low temperatures. Jayaprakash et al. (2023) employed electrostatic spray drying (ESD) as a novel method to encapsulate β-galactosidase. Their study shown that ESD effectively preserved the activity of β- galactosidase by employing different matrices. Therefore, the enclosed β-galactosidase has the ability to generate lactose-free or low-lactose foods while preserving its original enzymatic activity. Different methods of encapsulation are detailed in Fig. 4.2 and studies on utilization of different immobilization techniques for lactose hydrolysis are given in Table 4.1.

4.3.5 Membrane Reactors and Filtration

Lactose hydrolysis in dairy products can take place in various systems and configurations. Two often utilized types of reactors are compacted bed reactors (Fischer et al., 2013) and membrane reactors (Vasileva et al., 2016). A rotating horizontal column reactor is a widely adapted one at an industrial scale. By 1977, an industrial facility successfully processed 10 tons of milk per day. The durability of lactase

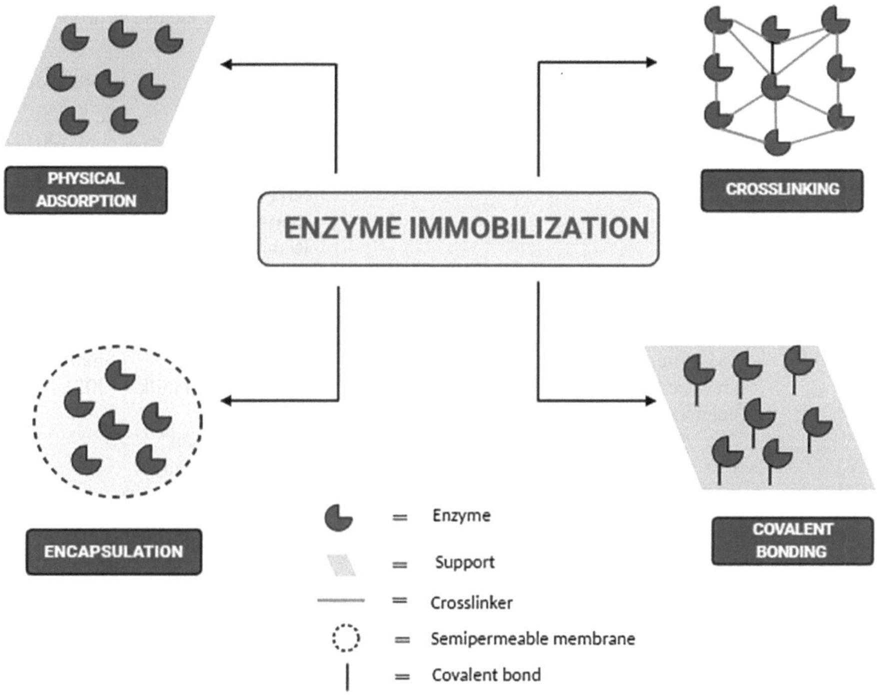

Fig. 4.2 Methods of immobilization techniques

reactors is influenced by factors such as temperature, substrate flow, carrier material, immobilization method, and substrate components (Finocchiaro et al., 1980). Subsequently, an immobilization technology offered the re-use of an enzyme up to 50 cycles with minimal loss of enzyme activity (<9%) as reported by Basso and Serban (2019). Immobilized enzymes are well-suited for usage in continuous systems and possess the advantage of being easily separated from the product for reuse or recycling. This simplicity enables the design of continuously running reactors with significant benefits. Hydrolysis is often conducted in a membrane reactor method using a protein-free liquid, such as milk or whey ultrafiltration permeate. The enzyme is extracted from the reaction mixture using a secondary ultrafiltration (UF) device, while the permeate, which no longer contains the enzyme, is combined again with the milk or whey retentate, together with the hydrolyzed lactose (Palai et al., 2014). The primary forms of enzyme reactors utilized for lactose hydrolysis are enzyme-membrane systems, fluidized-bed reactors, hollow-fiber reactors, packed-bed reactors, and stirred-tank reactors (Agyei et al., 2015).

In a membrane reactor (MBR), the enzyme is contained inside a specific area of space by the use of a selective membrane. Alternatively, the enzyme can be immobilized by being adsorbed or trapped within the membrane itself. The process of MBR for lactose or whey hydrolysis typically entails utilizing a thermostatic

Table 4.1 Studies on utilization of different immobilization techniques for lactose hydrolysis

Method	Support	Source of β-galactosidase	Findings	References
Physical adsorption	Sodium alginate or l-carrageenan	*Kluyveromyces lactis*	Yield of these systems is greater than 95% at pH 4.0 No effect on thermal stability Reduction in the denaturation rate of the enzyme exposed to low pH	de Souza et al. (2017)
	Modified nanosilver reduced graphene oxide nanocomposite	*Aspergillus oryzae*	89% of lactose hydrolysed in milk whey 85% of its total activity retained even after its 10 repeated applications	Shafi et al. (2021)
	Fe-chelated cryogel disks.	*Aspergillus oryzae*	After 25-runs, 64.9% of activity was preserved with high operational stability % of hydrolysed lactose is 80.0 ± 2.0 for 24 h	Inanan, (2022)
Cross-linking	Arabic gum-based hydrogel	*Kluyveromyces lactis*	Hydrolysis of standard lactose and lactose contained in UHT milk was 52.79 ± 0.85 and 93.92 ± 1.05 wt-% respectively Immobilized one could be employed for at least three successive cycles for the production of lactose-free/ low-lactose foods	Wolf et al. (2018)

(continued)

Table 4.1 (continued)

Method	Support	Source of β-galactosidase	Findings	References
Crosslinking and covalent bonding	Chitosan activated with glutaraldehyde, epichlorohydrin or glycidol	*Kluyveromyces lactis* NRRL Y1564	The optimum temperature and pH for the hydrolysis of lactose using the immobilized enzyme was 40 °C and 7.0, respectively Lactose hydrolysis in milk was 42.8% whereas in pre-treated whey, it was 86%	Vidal et al. (2020)
	Pectin-based hydrogel and pectin-based composite hydrogel	*Kluyveromyces lactis*	Enzymatic activity of immobilised enzyme was 86% greater than that determined for the free enzyme Retained for at least 6 successive cycles without completely losing enzymatic activity	Hanauer and Paulino (2022)
	Conjugation of polydopamine (PDA) coated magnetic particles grafted with diaminopolyethylene glycol (DAPEG) without using any activating agent.	*Aspergillus oryzae*	High pH tolerance and the optimum temperature was 15 °C higher than that of the free β-galactosidase 20 repeated cycles of lactose hydrolysis with good stability and slight loss in its activity	Bayramoglu et al. (2023)

(continued)

Table 4.1 (continued)

Method	Support	Source of β-galactosidase	Findings	References
Encapsulation	Encapsulated in solid/oil/water emulsions	(Maxilact® LGX5000F, in liquid form) Commercial grade lactase	Lactose hydrolysis in full-fat or skim milk reduced from >75% for free lactase to <15% for encapsulated lactase after 3-week storage	Zhang and Zhong (2018)
	Silica microcapsules	*Streptococcus lactis*	Optimum temperature of 45 °C and pH of 7.0 2.7 times more thermostable than the free cells	Mukundan et al. (2020)
	Halloysite nanotubes adsorbed enzyme entrapped by cellulose nanocrystals matrix	*Bifidobacterium bifidum*	Retained approximately 76% activity after 12 cycles of repeated use More thermostable at 55 °C than the free enzyme	Tizchang et al. (2021)

stirring vessel to allow for hydrolysis. The contents of the vessel are constantly circulated towards an ultrafiltration module. Utilizing MBR as a continuous method for lactose hydrolysis in whole milk or cheese whey is a highly efficient technology. Achieving both lactose conversion and recovery of high molecular weight proteins can be done simultaneously in a single step (Giorno & Drioli, 2000). The primary benefit of this arrangement is the ability to operate with unbound enzyme in uniform solutions alongside its substrate, so fully harnessing its inherent kinetic characteristics. The primary drawbacks of this setup include the increased susceptibility to microbial contamination, particularly when operating for extended periods at room temperatures, as well as the obstruction of ultra-filtration membranes by milk proteins. To mitigate these limitations, one can operate the system at elevated temperatures and employ deproteinated substrates such whey permeate (Husain, 2010). The continuous stirred tank reactor (CSTR) is a system in which reactants are continually fed into a well-stirred tank, maintaining a constant flow and equilibrium condition. Connecting CSTRs in series offers significant advantages due to the high level of mixing in each vessel. This ensures that the reactor volume is fully utilized for the reaction, without any unused space (Rouf et al., 2017). The CSTR is commonly employed in large-scale industrial production. However, for smaller-scale operations, batch processing is favored due to its superior production capabilities. Furthermore, the uninterrupted functioning leads to increased product stability, improved energy efficiency, and enhanced productivity.

The immobilized cell reactor known as a packed bed reactor is widely utilized in industry (Katchalski-Katzir, 1993). The design of these reactors is simple, comprising a column filled with biocatalysts that are constantly flushed by the liquid phase. Therefore, a wide range of materials, including delicate ones, can be employed for cell adsorption in packed bed reactors (Willaert et al., 2004). The use of cotton fibre as the carrier offers the advantages of being cost-effective and having a high surface area per unit mass. Additionally, the immobilization technique is simple and does not require expensive equipment. Therefore, this approach can be readily utilized on an industrial level for the extensive hydrolysis of lactose.

Plug flow bioreactors (PBRs) exhibit a higher average reaction rate in comparison to continuous stirred tank reactors. This is attributed to two factors: firstly, the inhibition effect is diminished due to the minimal disparity between substrate and product concentrations in the reactor. Secondly, enzyme loss is minimized as there are no collisions between biocatalyst particles and impeller, thereby reducing liquid shearing (Husain, 2010).

A fluidized bed reactor (FBR) uses a vessel with a porous plate at the bottom through which hot air or gas travels, and the catalysts are placed above the porous plate. As the velocity of the gas increases, the bed undergoes fluidization, causing an expansion in volume and rendering the gas bubbles invisible. In contrast to fixed bed reactors, fluidized bed reactors enable the introduction of more catalyst from the top of the reactor (Eş et al., 2015). The exceptional heat transmission is regarded as a primary advantage of this reactor design, enabling secure and efficient operation even for extremely exothermic reactions (Andreas Seidel-Morgenstern, 2010). The process of enzymatic hydrolysis using β-galactosidase immobilized on cellulose beads, known as FBR, was employed to break down lactose found in milk whey and whole milk (Roy & Gupta, 2003). However, the conversion of lactose in whole milk was limited to only 60% due to the presence of fat, which hinders the effectiveness of the FBR. FBR is ideal for handling viscous or particle substrates due to their ability to minimize pressure loss and ensure a uniform flow distribution over the axial section of the reactor. In addition, mechanical agitation is avoided and the support material remains undamaged upon contact with the impeller.

A continuous enzymatic bioconversion incurs further expenses for enzyme recovery and immobilization at the end of the process. The enzymatic membrane reactor (EMR) is a distinctive reactor design that enables the simultaneous facilitation of catalytic reaction and product separation, particularly for the hydrolysis of large molecules. Thus, a more economically viable procedure can be achieved. The term EMR refers to a biological reaction that is catalyzed by specific enzyme(s) or enzyme-producing cells and is combined with membrane separation. According to Sitanggang et al. (2022) EMR has demonstrated the capacity to enhance the efficiency of enzyme-promoted bioconversion, elevate product yield, and may be readily expanded for industrial applications in comparison to traditional reactors. GOS have been produced in EMRs through the utilization of soluble biolactate N5, which is an enzymatic product derived from Bacillus circulans (Cao et al., 2020). The study examined the steady-state performance of the EMR by altering the residence time (1.1–2.8 h) and enzyme load (17–190 U/g), while keeping operational settings constant, including temperature (50 °C), pH (6.0), lactose feed content (300 g/kg), and recirculation flow-rate (0.18 m3/h). While the stirred tank reactor produced a higher yield of glucose oxidase (GOS) with a higher degree of polymerization (3–6) at approximately 38% of the total carbohydrate basis, the enzymatic membrane reactor (EMR) showed consistent catalytic performance for over 120 h without any significant decline in product quality.

4.4 Fermentation

In the nineteenth century, experts directed their attention towards the removal of lactose from milk through the process of fermentation utilizing unbound cells of *S. fragilis* (Fermented milk is completely devoid of sugar and does not undergo significant changes in its nutritional composition. It is suitable for consumption by those with lactose intolerance, as well as those with diabetes and galactosemia (Rao et al., 1998). The process of lactic acid fermentation by bacteria is responsible for the breakdown of lactose, galactose, and glucose, which are digested by microflora. Consequently, long-fermented products, such as aged hard cheese, have extremely low levels of all sugars. Fermentation breaks down approximately 20–30% of the lactose in milk, as well as proteins and fats, into peptides, amino acids, and fatty acids. This process enhances the digestibility and absorption of nutrients for humans. Filamentous fungus frequently metabolizes lactose at extremely low rates (Li et al., 2023). Fungi have two main options for breaking down lactose: (1) breaking it down outside the cell and then absorbing the resultant glucose and galactose molecules, or (2) absorbing the lactose intact and then breaking it down inside the cell. Fungi species like *Aspergillus nidulans*, *Neurospora crassa,* and *Fusarium graminearum* employ the second approach for lactose utilization. On the other hand, fungi such as *Hypocrea jecorina (Trichoderma reesei)* and *Aspergillus niger* possess the capability to excrete β-galactosidase, which breaks down lactose in the external environment. The aerobic assimilation of lactose is a common trait among yeasts; however, the fermentation of lactose is very uncommon. Examples of yeasts that can ferment lactose include *Kluyveromyces lactis, K. marxianus, and Candida pseudotropicalis* (Seiboth et al., 2007). The hydrolysis of lactose can be carried out concurrently with fermentation. The technique known as co-hydrolysis is typically favored due to its ability to save time and eliminate the need for additional energy in various products (Schmidt et al., 2016). The process of lactose hydrolysis in milk using β-galactosidase to produce yogurt can be conducted either prior to fermentation, with varying time and temperature conditions, or during the fermentation process itself (Tamime & Robinson, 2007). There is a limited number of articles that discuss various aspects of lactose hydrolyzed yogurts. Inconsistent findings have been observed relating the lactose concentration and the physical and sensory attributes of the end product. The influence of this fact is dependent upon the specific enzyme used, the timing of its addition (before to fermentation or alongside the starter), and the storage conditions of the yogurt. *Lactobacillus bulgaricus* and *Streptococcus thermophilus* exhibit greater β-galactosidase production compared to other bacteria with lactase capabilities. Approximately 40–50% of the lactose is broken down during fermentation. The remaining amount of lactose, along with the presence of high levels of lactase activity, makes yogurt an ideal product for those with lactase deficiency. These individuals can absorb around 50% of a lactose load (Kilara & Shahani, 1976). In their study, Nagaraj et al. (2009), investigated several concentrations of lactase enzyme (Lactozyme®) to determine the optimal quantity of lactose hydrolysis required for the production of high-quality yoghurt. An observation was made that when the enzyme concentration grew from 0.12 to 0.56%, the extent of lactose hydrolysis (LH) also increased from 46.3 to 90.06%. Yoghurt of high quality can be made by using a mixture of lactose hydrolysed yoghurt that contains between 50% and 70% lactose. This method offers improved technology for producing enjoyable yogurt with a reduced lactose content, and it also appeals to a larger spectrum of consumers. Cheese is a dairy product with naturally low levels of lactose. The majority of lactose will be eliminated through whey during the processing and production of cheese. During the ripening process of hard and semihard cheese, the lactose that remains is further transformed into lactic acid, resulting in the production of natural lactose-free products. The LAB commonly employed as starter cultures for Camembert manufacturing include *Lactococcus lactis subsp. lactis*, *Lactococcus lactis subsp. cremoris*, and rarely *Streptococcus thermophilus*. They vary in the metabolic pathway employed for lactose metabolism. Both *Lactococcus lactis subsp.* translocate lactose using a phosphotransferase mechanism. Lactose is transported into the cytoplasm of these organisms in the form of lactose-phosphate and is then broken down by phospho-β-galactosidase (Ray et al., 2023). Lactose is able to enter the *Streptococcus thermophilus* cell through the assistance of lactose permease. The initial stage in the cytoplasmic cell is the process of hydrolysis, facilitated by β-galactosidase. Glucose is metabolized via the glycolytic pathway to produce L-lactate, whereas galactose is secreted into the medium (cheese). When the media contains only a limited proportion of lactose, strains of *Streptococcus thermophilus* that are capable of fermenting galactose through the Leloir route can do so. Panseri et al. (2021) identified the presence of carbohydrates in dairy products that are devoid of lactose. Furthermore, the lactose concentration in PDO (Protected Designation of Origin) hard cheese was found to be less than 0.0001 mg/kg. There is a diverse range of low-lactose/lactose-free cheeses readily accessible in the market to cater to lactose-intolerant individuals.

The lactose levels in Asiago PDO, Gorgonzola PDO, Emmentaler PDO, Pecorino Toscano PDO, Piave PDO, Stelvio PDO, and Montasio PDO vary from 100–1000 mg/kg (Facioni et al., 2021). The lactose concentration in Grana Padano PDO, Parmigiano Reggiano PDO, and Pecorino Romano PDO is below 10 mg/kg. Lactose hydrolysis enables the production of a novel variety of frozen yogurt that is suitable for individuals with lactose sensitivity. In addition, the process of breaking down lactose into monosaccharides in yogurt lowers its freezing point. This leads to a significant improvement in the texture and thickness of the product. It also helps prevent ice crystals from forming again and may slow down the rate at which the yogurt melts, even without the use of stabilizers (Skryplonek et al., 2017).

4.5 Conclusion

The market expansion of hydrolysed lactose appears to be propelled by a rising recognition of the issue of lactose intolerance and shifting demographic patterns. In this approach, enzymatic immobilization of β-galactosidase is a cost-effective method for lactose hydrolysis, however it is afflicted by significant issues such as microbial contamination, protein adhesion, and channelling. Implementing regular washing and pasteurization, as well as adjusting the flow direction of the feed, can effectively address these issues. The hydrolysis of lactose in dairy products can take place in many methods and configurations include compacted bed reactors and membrane reactors, which are among the most commonly utilized types of reactors. EMRs possess the potential to be employed in sustainable and ecologically conscious biocatalytic processes. Moreover, they have the capacity to augment the efficacy and durability of biocatalytic reactions, rendering them highly appealing competitors for green chemical applications (Table 4.1).

References

Abd-El-Salam, M. H., El-Shibiny, S., Ahmed, N. S., & Ismail, A. A. (2020). Use of ultrafiltration in manufacture of Domiati cheese from buffalo milk. *Egyptian Journal of Dairy Science, 9*, 151–157.

Agyei, D., Shanbhag, B. K., & He, L. (2015). Enzyme engineering (immobilization) for food applications. In *Improving and tailoring enzymes for food quality and functionality* (pp. 213–235). Woodhead Publishing. https://doi.org/10.1016/B978-1-78242-285-3.00011-9

Ahn, S. I., Lee, Y. K., & Kwak, H. S. (2019). Physicochemical and sensory properties of milk supplemented with lactase microcapsules coated with enteric coating materials. *Journal of Dairy Science, 102*(8), 6959–6970. https://doi.org/10.3168/jds.2018-15865

Ansari, S. A., & Husain, Q. (2012). Lactose hydrolysis from milk/whey in batch and continuous processes by concanavalin A-Celite 545 immobilized Aspergillus oryzae β galactosidase. *Food and Bioproducts Processing, 90*(2), 351–359. https://doi.org/10.1016/j.fbp.2011.07.003

Aronson, M. (1952). Transgalactosidation during lactose hydrolysis. *Archives of Biochemistry and Biophysics, 39*(2), 370–378. https://doi.org/10.1016/0003-9861(52)90346-9

Basso, A., & Serban, S. (2019). Industrial applications of immobilized enzymes—A review. *Molecular Catalysis, 479*, 110607. https://doi.org/10.1016/j.mcat.2019.110607

Bayramoglu, G., Cimen, A. G., & Arica, M. Y. (2023). Immobilisation of β-galactosidase onto double layered hydrophilic polymer coated magnetic nanoparticles: Preparation, characterisation and lactose hydrolysis. *International Dairy Journal, 138*, 105545. https://doi.org/10.1016/j.idairyj.2022.105545

Bhardwaj, T. R., Kanwar, M., Lal, R., & Gupta, A. (2000). Natural gums and modified natural gums as sustained-release carriers. *Drug Development and Industrial Pharmacy, 26*(10), 1025–1038. https://doi.org/10.1081/DDC-100100266

Cao, T., Pázmándi, M., Galambos, I., & Kovács, Z. (2020). Continuous production of galacto-oligosaccharides by an enzyme membrane reactor utilizing free enzymes. *Membranes, 10*(9), 1–16. https://doi.org/10.3390/membranes10090203

Cargnin, M. A., Gasparin, B. C., dos Santos Rosa, D., & Paulino, A. T. (2021). Performance of lactase encapsulated in pectin-based hydrogels during lactose hydrolysis reactions. *Lwt, 150*, 111863. https://doi.org/10.1016/j.lwt.2021.111863

Cheng, S., & Martínez-Monteagudo, S. I. (2019). Hydrogenation of lactose for the production of lactitol. *Asia-Pacific Journal of Chemical Engineering, 14*(1), 2275. https://doi.org/10.1002/apj.2275

Coughlin, J. R., & Nickerson, T. A. (1975). Acid-Catalyzed Hydrolysis of Lactose in Whey and Aqueous Solutions. *Journal of Dairy Science, 58*(2), 169–174. https://doi.org/10.3168/jds.S0022-0302(75)84540-1

Dahlqvist, A., Asp, N. G., Burvall, A., & Rausing, H. (1977). Hydrolysis of lactose in milk and whey with minute amounts of lactase. *Journal of Dairy Research, 44*(3), 541–548. https://doi.org/10.1017/S0022029900020495

Damin, B. I. S., Kovalski, F. C., Fischer, J., Piccin, J. S., & Dettmer, A. (2021). Challenges and perspectives of the β-galactosidase enzyme. *Applied Microbiology and Biotechnology, 105*(13), 5281–5298. https://doi.org/10.1007/s00253-021-11423-7

de Souza, L. T. A., Veríssimo, L. A. A., João, B. C. P., Santoro, M. M., Resende, R. R., & Mendes, A. A. (2017). Imobilização enzimática: princípios fundamentais e tipos de suporte. *Biotecnologia Aplicada à Agro&Indústria, 4*(4), 529–568. https://doi.org/10.5151/9788521211150-15

Dekker, P. J., Koenders, D., & Bruins, M. J. (2019). Lactose-free dairy products: Market developments, production, nutrition and health benefits. *Nutrients, 11*(3), 551.

Demirhan, E., Apar, D. K., & Özbek, B. (2010). A modelling study on hydrolysis of whey lactose and stability of β-galactosidase. *Korean Journal of Chemical Engineering, 27*(2), 536–545. https://doi.org/10.2478/s11814-010-0062-5

Deng, Z., Deng, Q., Li, B., Li, J., Jung, S., Cho, N. J., & Liang, H. (2024). Strategies for lactase immobilization and delivery to relieve lactose intolerance. *Trends in Food Science and Technology, 143*, 104244. https://doi.org/10.1016/j.tifs.2023.104244

Duan, F., Sun, T., Zhang, J., Wang, K., Wen, Y., & Lu, L. (2022). Recent innovations in immobilization of β-galactosidases for industrial and therapeutic applications. *Biotechnology Advances, 61*, 108053. https://doi.org/10.1016/j.biotechadv.2022.108053

Elnashar, M. M. M., & Yassin, M. A. (2009). Covalent immobilization of β-galactosidase on carrageenan coated with chitosan. *Journal of Applied Polymer Science, 114*(1), 17–24. https://doi.org/10.1002/app.30535

Eş, I., Vieira, J. D. G., & Amaral, A. C. (2015). Principles, techniques, and applications of biocatalyst immobilization for industrial application. *Applied Microbiology and Biotechnology, 99*(5), 2065–2082. https://doi.org/10.1007/s00253-015-6390-y

Estevinho, B. N., Damas, A. M., Martins, P., & Rocha, F. (2014). Microencapsulation of β-galactosidase with different biopolymers by a spray-drying process. *Food Research International, 64*, 134–140. https://doi.org/10.1016/j.foodres.2014.05.057

Facioni, M. S., Dominici, S., Marescotti, F., Covucci, R., Taglieri, I., Venturi, F., & Zinnai, A. (2021). Lactose residual content in pdo cheeses: Novel inclusions for consumers with lactose intolerance. *Food, 10*(9), 2236. https://doi.org/10.3390/foods10092236

Finocchiaro, T., Olson, N. F., & Richardson, T. (1980). Use of immobilized lactase in milk systems. In *Advances in biochemical engineering* (Vol. 15). Springer. https://doi.org/10.1007/3540096868_3

Fischer, C., & Kleinschmidt, T. (2018). Synthesis of galactooligosaccharides in milk and whey: A review. *Comprehensive Reviews in Food Science and Food Safety, 17*(3), 678–697. https://doi.org/10.1111/1541-4337.12344

Fischer, J., Guidini, C. Z., Santana, L. N. S., De Resende, M. M., Cardoso, V. L., & Ribeiro, E. J. (2013). Optimization and modeling of lactose hydrolysis in a packed bed system using immobilized β-galactosidase from Aspergillus oryzae. *Journal of Molecular Catalysis B: Enzymatic, 85–86*, 178–186. https://doi.org/10.1016/j.molcatb.2012.09.008

Giorno, L., & Drioli, E. (2000). Biocatalytic membrane reactors: Applications and perspectives. *Trends in Biotechnology, 18*(8), 339–349. https://doi.org/10.1016/S0167-7799(00)01472-4

Gosling, A., Stevens, G. W., Barber, A. R., Kentish, S. E., & Gras, S. L. (2010). Recent advances refining galactooligosaccharide production from lactose. *Food Chemistry, 121*(2), 307–318.

Haider, T., & Husain, Q. (2007). Preparation of lactose-free milk by using salt-fractionated almond (Amygdalus communis) β-galactosidase. *Journal of the Science of Food and Agriculture, 87*(7), 1278–1283. https://doi.org/10.1002/jsfa.2840

Hanauer, D. C., & Paulino, A. T. (2022). Anchoring lactase in pectin-based hydrogels for lactose hydrolysis reactions. *Process Biochemistry, 122*, 50–59. https://doi.org/10.1016/j.procbio.2022.08.026

Harju, M. (2004). Chromatographic and enzymatic removal of lactose from milk. *Bulletin-International Dairy Federation, 3898*, 4–8. http://cat.inist.fr/?aModele=afficheN&cpsidt=15840290

Harju, M., Kallioinen, H., & Tossavainen, O. (2012). Lactose hydrolysis and other conversions in dairy products: Technological aspects. *International Dairy Journal, 22*(2), 104–109. https://doi.org/10.1016/j.idairyj.2011.09.011

Hartmeier, W., Business Media, Gekas, V., & Lopez-Leiva, M. (2012). Immobilized biocatalysts: An introduction. *Process Biochemistry, 20*(1).

Huh, K. T., Toba, T., & Adachi, S. (1991). Oligosaccharide structures formed during acid hydrolysis of lactose. *Food Chemistry, 39*(1), 39–49. https://doi.org/10.1016/0308-8146(91)90083-Z

Husain, Q. (2010). β Galactosidases and their potential applications: A review. *Critical Reviews in Biotechnology, 30*(1), 41–62. https://doi.org/10.3109/07388550903330497

Inanan, T. (2022). Cryogel disks for lactase immobilization and lactose-free milk production. *Lwt, 154*, 112608. https://doi.org/10.1016/j.lwt.2021.112608

Jayaprakash, P., Gaiani, C., Edorh, J. M., Beaupeux, E., Maudhuit, A., & Desobry, S. (2023). Impact of matrices composition and processes on β-galactosidase encapsulation. *Journal of Food Engineering, 353*, 111547. https://doi.org/10.1016/j.jfoodeng.2023.111547

Jelen, P., & Tossavainen, O. (2003). Low lactose and lactose-free milk and dairy products – Prospects, technologies and applications. *Australian Journal of Dairy Technology, 58*(2), 161–165.

Jørgensen, F., Hansen, O. C., & Stougaard, P. (2001). High-efficiency synthesis of oligosaccharides with a truncated β-galactosidase from Bifidobacterium bifidum. *Applied Microbiology and Biotechnology, 57*(5–6), 647–652. https://doi.org/10.1007/s00253-001-0845-z

Kalathinathan, P., Sain, A., Pulicherla, K., & Kodiveri Muthukaliannan, G. (2023). A Review on the Various Sources of β-Galactosidase and Its Lactose Hydrolysis Property. *Current Microbiology, 80*(4), 122. https://doi.org/10.1007/s00284-023-03220-4

Katchalski-Katzir, E. (1993). Immobilized enzymes – learning from past successes and failures. *Trends in Biotechnology, 11*(11), 471–478. https://doi.org/10.1016/0167-7799(93)90080-S

Katrolia, P., Liu, X., Li, G., & Kopparapu, N. K. (2019). Enhanced Properties and Lactose Hydrolysis Efficiencies of Food-Grade β-Galactosidases Immobilized on Various Supports: a Comparative Approach. *Applied Biochemistry and Biotechnology, 188*(2), 410–423. https://doi.org/10.1007/s12010-018-2927-8

Kilara, A., & Shahani, K. M. (1976). Lactase activity of cultured and acidified dairy products. *Journal of Dairy Science, 59*(12), 2031–2035.

Klein, M. P., Hackenhaar, C. R., Lorenzoni, A. S. G., Rodrigues, R. C., Costa, T. M. H., Ninow, J. L., & Hertz, P. F. (2016). Chitosan crosslinked with genipin as support matrix for application in food process: Support characterization and β-d-galactosidase immobilization. *Carbohydrate Polymers, 137*, 184–190. https://doi.org/10.1016/j.carbpol.2015.10.069

Li, A., Zheng, J., Han, X., Yang, S., Cheng, S., Zhao, J., Zhou, W., & Lu, Y. (2023). Advances in low-lactose/lactose-free dairy products and their production. *Food, 12*(13), 2553. https://doi.org/10.3390/foods12132553

Lv, Y., & Yu, G. (2014). Galactose: Chemical structure, dietary sources and clinical significance. *Galactose: Structure and Function in Biology and Medicine, 0*, 1–23.

Macwan, S. R., Dabhi, B. K., Parmar, S. C., & Aparnathi, K. D. (2016). Whey and its Utilization. *International Journal of Current Microbiology and Applied Sciences, 5*(8), 134–155. https://doi.org/10.20546/ijcmas.2016.508.016

Malecki, E., Viere, R., & Rosemeyer, H. (2013). Immobilization of 5-fluorouridine on chitosan. *Chemistry and Biodiversity, 10*(10), 1828–1841. https://doi.org/10.1002/cbdv.201300025

Mukundan, S., Melo, J. S., Sen, D., & Bahadur, J. (2020). Enhancement in β-galactosidase activity of Streptococcus lactis cells by entrapping in microcapsules comprising of correlated silica nanoparticles. *Colloids and Surfaces B: Biointerfaces, 195*, 111245. https://doi.org/10.1016/j.colsurfb.2020.111245

Nagaraj, M., Sharanagouda, B., Manjunath, H., & Manafi, M. (2009). Standardization of different levels of lactose hydrolysis in the preparation of lactose hydrolyzed yoghurt. *Iranian Journal of Veterinary Research, 10*(2), 132–136.

Nath, A., Bhattacharjee, C., & Chowdhury, R. (2013). Synthesis and separation of galacto-oligosaccharides using membrane bioreactor. *Desalination, 316*, 31–41. https://doi.org/10.1016/j.desal.2013.01.024

Nath, A., Mondal, S., Chakraborty, S., Bhattacharjee, C., & Chowdhury, R. (2014). Production, purification, characterization, immobilization, and application of β-galactosidase: a review. *Asia-Pacific Journal of Chemical Engineering, 9*(3), 330–348.

Ney, K. H., & Wirotama, I. P. G. (1970). Investigation of lactose hydrolysis. *Zeitschrift Für Lebensmittel-Untersuchung Und-Forschung, 143*, 93–95.

Oliveira, R. C., Ribeiro, L. B., de Souza, T. C., de Freitas, L. A., de Almeida, A. C. P., & Gonçalves, L. R. B. (2023). Carbohydrate-active enzymes in the production of lactose-derived tagatose. In *Polysaccharide degrading biocatalysts* (pp. 365–383). Academic. https://doi.org/10.1016/B978-0-323-99986-1.00008-9

Paiva, M. J., Paula-Elias, F. C., Pereira, L. A., Carreiro, S. C., Vieira-Almeida, E. C., Silva, E. M., Dias, G. S., Xavier, M. A. C., Morales, S. A., Perna, R. F., & Almeida, A. F. (2023). Stabilization of β-Galactosidase Encapsulated in Pectin-Alginate Hydrogel and Hydrolysis of Whey Lactose and Whole Milk. In. *Journal of the Brazilian Chemical Society, 34*(12), 1743–1752. https://doi.org/10.21577/0103-5053.20230067

Palai, T., Singh, A. K., & Bhattacharya, P. K. (2014). Enzyme, β-galactosidase immobilized on membrane surface for galacto-oligosaccharides formation from lactose: Kinetic study with feed flow under recirculation loop. *Biochemical Engineering Journal, 88*, 68–76. https://doi.org/10.1016/j.bej.2014.03.017

Panesar, P. S., Panesar, R., Singh, R. S., Kennedy, J. F., & Kumar, H. (2006). Microbial production, immobilization and applications of β-D-galactosidase. *Journal of Chemical Technology and Biotechnology, 81*(4), 530–543. https://doi.org/10.1002/jctb.1453

Panesar, P. S., Kumari, S., & Panesar, R. (2010). Potential applications of immobilized β-galactosidase in food processing industries. *Enzyme Research, 2010*, 1–16. https://doi.org/10.4061/2010/473137

Panseri, S., Pavlovic, R., Castrica, M., Nobile, M., Di Cesare, F., & Chiesa, L. M. (2021). Determination of carbohydrates in lactose-free dairy products to support food labelling. *Food, 10*(6), 1219. https://doi.org/10.3390/foods10061219

Park, A. R., & Oh, D. K. (2010). Galacto-oligosaccharide production using microbial β-galactosidase: Current state and perspectives. *Applied Microbiology and Biotechnology, 85*(5), 1279–1286. https://doi.org/10.1007/s00253-009-2356-2

Perotti, M. C., Veronica, W. I., Ines, V. C., & Viviana, B. C. (2012). Dairy products modified in their lactose content. *Current Nutrition & Food Science, 8*(1), 8–18.

Popescu, L., Bulgaru, V., & Siminiuc, R. (2021). Effect of temperature, pH and amount of enzyme used in the lactose hydrolysis of milk. *Food and Nutrition Sciences, 12*(12), 1243–1254. https://doi.org/10.4236/fns.2021.1212091

Rabiu, B. A., Jay, A. J., Gibson, G. R., & Rastall, R. A. (2001). Synthesis and fermentation properties of novel galacto-oligosaccharides by β-Galactosidases from bifidobacterium species. *Applied and Environmental Microbiology, 67*(6), 2526–2530. https://doi.org/10.1128/AEM.67.6.2526-2530.2001

Ramsdell, G. A., & Webb, B. H. (1945). The acid hydrolysis of lactose and the preparation of hydrolyzed lactose sirup. *Journal of Dairy Science, 28*(9), 677–686. https://doi.org/10.3168/jds.S0022-0302(45)95222-2

Rao, B. Y. K., Godbole, S. S., & De Souz, S. (1998). Preparation of lactose free milk by fermentation using immobilized Saccharomyces fragilis. *Journal of Food Safety & Quality, 13*(18), 427–430.

Ray, M., Manjunath, A., Baruah, R., & Halami, P. M. (2023). Enzymes from lactic acid bacteria for nutraceuticals production. In *Microbial enzymes in production of functional foods and nutraceuticals* (pp. 25–44). CRC Press.

Rico-Rodríguez, F., Villamiel, M., Ruiz-Aceituno, L., Serrato, J. C., & Montilla, A. (2020). Effect of the lactose source on the ultrasound-assisted enzymatic production of galactooligosaccharides and gluconic acid. *Ultrasonics Sonochemistry, 67*, 104945.

Rouf, A., Kanojia, V., & Naik, H. R. (2017). Cell immobilization: An overview on techniques and its applications in food industry. *International Journal of Chemical Studies, 5*(6), 1817–1824.

Roy, I., & Gupta, M. N. (2003). Lactose hydrolysis by Lactozym ™ immobilized on cellulose beads in batch and fluidized bed modes. *Process Biochemistry, 39*(3), 325–332. https://doi.org/10.1016/S0032-9592(03)00086-4

Schaafsma, G. (2008). Lactose and lactose derivatives as bioactive ingredients in human nutrition. *International Dairy Journal, 18*(5), 458–465. https://doi.org/10.1016/j.idairyj.2007.11.013

Schmidt, C., Mende, S., Jaros, D., & Rohm, H. (2016). Fermented milk products: effects of lactose hydrolysis and fermentation conditions on the rheological properties. *Dairy Science and Technology, 96*(2), 199–211. https://doi.org/10.1007/s13594-015-0259-9

Seiboth, B., Pakdaman, B. S., Hartl, L., & Kubicek, C. P. (2007). Lactose metabolism in filamentous fungi: how to deal with an unknown substrate. *Fungal Biology Reviews, 21*(1), 42–48. https://doi.org/10.1016/j.fbr.2007.02.006

Seidel-Morgenstern, A. (2010). Membrane reactors: Distributing reactants to improve selectivity and yield. In A. Seidel-Morgenstern (Ed.), *Membrane reactors: Distributing reactants to improve selectivity and yield*. Wiley. https://doi.org/10.1002/9783527629725

Shafi, A., Ahmed, F., & Husain, Q. (2021). β-Galactosidase mediated synthesized nanosupport for the immobilization of same enzyme: Its stability and application in the hydrolysis of lactose. *International Journal of Biological Macromolecules, 184*, 57–67. https://doi.org/10.1016/j.ijbiomac.2021.06.034

Singh, R. V., & Sambyal, K. (2023). B-Galactosidase as an industrial enzyme: Production and potential. *Chemical Papers, 77*(1), 11–31. https://doi.org/10.1007/s11696-022-02507-3

Singh, P., Arora, S., Rao, P. S., Kathuria, D., Sharma, V., & Singh, A. K. (2021). Effect of process parameters on the β-galactosidase hydrolysis of lactose and galactooligosaccharide formation in concentrated skim milk. *Food Chemistry, 393*, 133355.

Sitanggang, A. B., Diaussie, K., Rosabel, C., & Budijanto, S. (2022). Current trends in enzymatic membrane reactor. In *Current trends and future developments on (Bio-) Membranes: Recent advances on membrane reactors* (pp. 195–225). https://doi.org/10.1016/B978-0-12-823659-8.00003-4

Skryplonek, K., Gomes, D., Viegas, J., Pereira, C., & Henriques, M. (2017). Lactose-free frozen yogurt: Production and characteristics. *Acta Scientiarum Polonorum Technologia Alimentaria, 16*(2), 171–179.

Souza, C. J. F., Garcia-Rojas, E. E., & Favaro-Trindade, C. S. (2018). Lactase (β-galactosidase) immobilization by complex formation: Impact of biopolymers on enzyme activity. *Food Hydrocolloids, 83*, 88–96. https://doi.org/10.1016/j.foodhyd.2018.04.044

Stanek, J., Cerney, M., & Pacak, J. (1965). Synthesis of oligosaccharides by the action of mineral acids: Reversion. In *The oligosaccharides* (pp. 108–115). Academic.

Tamime, A. Y., & Robinson, R. K. (2007). *Yoghurt: Science and technology*. Pergamon Press.

Tanriseven, A., & Doğan, Ş. (2002). A novel method for the immobilization of β-galactosidase. *Process Biochemistry, 38*(1), 27–30. https://doi.org/10.1016/S0032-9592(02)00049-3

Tizchang, S., Khiabani, M. S., Mokarram, R. R., Hamishehkar, H., Mohammadi, N. S., & Chisti, Y. (2021). Immobilization of β-galactosidase by halloysite-adsorption and entrapment in a cellulose nanocrystals matrix. *Biochimica et Biophysica Acta – General subjects, 1865*(6), 129896. https://doi.org/10.1016/j.bbagen.2021.129896

Troise, A. D., Bandini, E., De Donno, R., Meijer, G., Trezzi, M., & Fogliano, V. (2016). The quality of low lactose milk is affected by the side proteolytic activity of the lactase used in the production process. *Food Research International, 89*, 514–525. https://doi.org/10.1016/j.foodres.2016.08.021

Ugidos-Rodríguez, S., Matallana-González, M. C., & Sánchez-Mata, M. C. (2018). Lactose malabsorption and intolerance: a review. *Food and Function, 9*(8), 4056–4068. https://doi.org/10.1039/c8fo00555a

Ureta, M. M., Martins, G. N., Figueira, O., Pires, P. F., Castilho, P. C., & Gomez-Zavaglia, A. (2021). Recent advances in β-galactosidase and fructosyltransferase immobilization technology. *Critical Reviews in Food Science and Nutrition, 61*(16), 2659–2690. https://doi.org/10.1080/10408398.2020.1783639

Urrutia, P., Bernal, C., Wilson, L., & Illanes, A. (2018). Use of chitosan heterofunctionality for enzyme immobilization: β-galactosidase immobilization for galacto-oligosaccharide synthesis. *International Journal of Biological Macromolecules, 116*, 182–193. https://doi.org/10.1016/j.ijbiomac.2018.04.112

Vasileva, N., Ivanov, Y., Damyanova, S., Kostova, I., & Godjevargova, T. (2016). Hydrolysis of whey lactose by immobilized β-galactosidase in a bioreactor with a spirally wound membrane. *International Journal of Biological Macromolecules, 82*, 339–346. https://doi.org/10.1016/j.ijbiomac.2015.11.025

Vidal, V. A. S., Paglarini, C. S., Freitas, M. Q., Coimbra, L. O., Esmerino, E. A., Pollonio, M. A. R., & Cruz, A. G. (2020). Q Methodology: An interesting strategy for concept profile and sensory description of low sodium salted meat. *Meat Science, 161*, 108000. https://doi.org/10.1016/j.meatsci.2019.108000

Waghmare, R. B. (2020). Encapsulation techniques for delivery of bioactive compounds in milk and dairy products- A review. *Journal of Dairy Research & Technology, 3*(1), 1–9. https://doi.org/10.24966/drt-9315/100017

Wallenfels, K., & Malhotra, O. P. (1962). Galactosidases. In *Advances in carbohydrate chemistry* (Vol. 16, pp. 239–298). Academic.

Willaert, R., Nedović, V., & Baron, G. V. (2004). Physiology of immobilised microbial cells. In *Fundamentals of cell immobilisation biotechnology* (pp. 469–492). https://doi.org/10.1007/978-94-017-1638-3_24

Wolf, M., Gasparin, B. C., & Paulino, A. T. (2018). Hydrolysis of lactose using β-D-galactosidase immobilized in a modified Arabic gum-based hydrogel for the production of lactose-free/low-lactose milk. *International Journal of Biological Macromolecules, 115*, 157–164. https://doi.org/10.1016/j.ijbiomac.2018.04.058

Zadow, J. G. (1986). Lactose hydrolysed dairy products. *Food Technology in Australia, 38*(11), 460–471.

Zhang, Y., & Zhong, Q. (2018). Freeze-dried capsules prepared from emulsions with encapsulated lactase as a potential delivery system to control lactose hydrolysis in milk. *Food Chemistry, 241*, 397–402. https://doi.org/10.1016/j.foodchem.2017.09.004

Zhang, Z., Zhang, R., Chen, L., & McClements, D. J. (2016). Encapsulation of lactase (β-galactosidase) into κ-carrageenan-based hydrogel beads: Impact of environmental conditions on enzyme activity. *Food Chemistry, 200*, 69–75. https://doi.org/10.1016/j.foodchem.2016.01.014

Chapter 5
Effect of Processing Parameters on Lactose Hydrolysis

Ankita Hooda, Pranali Nikam, Abhishek Dutt Tripathi and Ananya Bharadwaj

5.1 Introduction

The lactose hydrolysis is a most promising process used in dairy industry for the production of lactose-free or reduced-lactose dairy products. By studying the parameters that influence lactose hydrolysis, researchers can develop more efficient methods to reduce lactose content in dairy products, making them suitable for consumption by lactose-intolerant individuals. These products cater to a growing market of consumers who prefer or require lactose-free options due to lactose intolerance, dietary choices, or other health considerations (Ansari & Husain, 2010, Freitas et al., 2011, Gerbault et al., 2011, Horner et al., 2011; Schulz & Rizvi, 2023). By studying the parameters, manufacturers can optimize the lactose hydrolysis process and ensure the production of high-quality, palatable, and safe lactose-reduced products. Lactose is a source of carbohydrates and energy in dairy products. By hydrolysing lactose, the resulting glucose and galactose can be more readily absorbed by the body. There are two common methods for the lactose hydrolysis: (a) Acid hydrolysis and (b) Enzymatic hydrolysis. In acid hydrolysis the reaction is rapid, as acids are act as a catalyst, but the temperature for the reaction is quite higher (150 °C), as a result the final quality of product is inferior due to incorporation of cooked flavour and colour. While in enzymatic hydrolysis, milk or whey are processed comparatively at lower temperature (30–40 °C) which prevents the nutritional attributes and improves the sweetness. Lactose hydrolysis by means of

A. Hooda · A. D. Tripathi · A. Bharadwaj
Department of Dairy Science and Food Technology, Banaras Hindu University, Varanasi, Uttar Pradesh, India

P. Nikam (✉)
Department of Dairy Chemistry, College of Dairy Science and Food Technology, Dau Shri Vasudev Chandrakar Kamdhenu Vishwavidyalaya, Raipur, Chhattisgarh, India

P. S. Rao et al. (eds.), *Lactose Hydrolysis in Dairy Products*,
https://doi.org/10.1007/978-3-031-78207-7_5

enzymatic process has various technological as well as environmental benefits of industrial applications. These can be:

1. Individuals with compromised lactose digestion, children, and the elderly, who may rely on dairy products for essential nutrients such as calcium and vitamin D can consume lactose hydrolysed dairy products.
2. Encourages the production of galactooligosaccharides (GOS), a non-digestible prebiotic (long chain of monomers from 2–8 galactose molecules containing one glucose molecule). GOS promotes the growth of intestinal microflora i.e. *Bifidobacteria* and *Lactobacill* (Schulz & Rizvi, 2023).
3. Allows better quality control of lactose-reduced dairy products, maintaining the desired sensory attributes and shelf life of the products such as enhanced solubility (lowers lactose crystallization and granny texture in condensed milk, milk powders and ice-reams), increased sweetness, reduced calories, increased softness and creaminess, easy to ferment (yogurt preparation) and minimized non-enzymatic Maillard reaction.
4. Improves the biodegradation of by-products (whey)

Overall, studying the parameters that influence lactose hydrolysis in dairy products is crucial for addressing consumer needs, improving product quality, ensuring nutritional benefits, optimizing processes, and contributing to sustainability efforts in the dairy industry.

5.2 Kinetics of Lactose Hydrolysis

The lactose hydrolysis by enzymatic process is catalysed by enzyme β-galactosidase (lactase) and can be obtained from various sources such as microbial, vegetal (plant) and animal. (see Chap. 3). The vegetal and animal sources are not found so common because these are expensive. The β-galactosidase obtained from bacteria i.e. *Escherichia coli* is mostly studied and enzyme obtained from *Escherichia coli* is not considered for use in free form in the dairy industry due to digestive malfunctioning. Enzymes obtained from yeasts and fungal resources are considered as safe and they have GRAS status. The enzymes from *Kluyveromyces* yeasts and *Aspergillus* fungi are mostly acceptable and commonly used for lactose hydrolysis in milk and dairy products. The mostly used β-galactosidase sources with their optimal experimental conditions are described in Table 5.1. The β-galactosidase obtained from fungal sources can be used in acidic whey (optimum pH range is 3.5–4.5) due to higher relative activity at low pH values. The β-galactosidase obtained from yeasts sources can be used in milk and rennet (sweet) whey (optimum pH range is 6.5–7.0).

For the lactose hydrolysis β-galactosidase enzyme are used either in free form or in immobilized form. For the industrial applications of enzyme for the commercial production of products, β-galactosidase is immobilized. Lactose hydrolysis mechanism by β-galactosidase enzyme has been studied by several researchers using different substrates under different experimental conditions. The kinetic studies

Table 5.1 Optimal experimental conditions for commonly used β-galactosidase

Source	Enzyme	Optimum pH	Optimum temperature (°C)
Bacterial	*Escherichia coli*	7.2	40
	Bacillus circulans	6.0	60
	Bacillus subtilis	6.5–7.0	50
	Bacillus sterarothermophilis	5.8–6.	65
	Lactobacillus acidophilus	6.2–6.6	55
	Streptococcus thermophilus	7.1	55
	L. delbrueckii subsp. bulgaricus and S. thermophilus	6.5	50
Fungal	*Aspergillus niger*	3.0–4.0	55–60
	Aspergillus oryzae	5.0	50–55
Yeast	*Kluyveromyces lactis*	6.9–7.3	35
	Kluyveromyces fragilis	6.6	37
	Kluyveromyces sp. PCH397 (Yak milk)	7.0	50
	Cryptococcus laurenti	4.57	55

(Source: Jurado et al., 2002; Schulz & Rizvi, 2023; Nag et al., 2021; Deshmukh et al., 2024)

proposed by several researchers for free (soluble) and immobilized enzyme used is shown in Table 5.2. The kinetic behaviour of immobilized enzymes than that of free ones may differs may because of conformational and diffusional effects. Various kinetic models for lactose hydrolysis are described in the literature by exploring the different β-galactosidase enzyme sources and different reaction mediums. The Michaelis–Menten kinetic model have been mostly used for fitting experimental data of lactose hydrolysis in which the substrate (lactose) and product (glucose and galactose) inhibition have been considered (Aliyah et al. (2017); Singh et al. (2021)). The various kinetic equations proposed by different workers for fitting the experimental data for lactose hydrolysis is given in Table 5.3.

5.3 Effect of Enzyme Action on Lactose Hydrolysis

β-galactosidase is the enzyme used to hydrolyse lactose into glucose and galactose. The generalized mechanism of β-galactosidase catalytic action on lactose involves a three steps leading to the hydrolysis as well as oligosaccharide formation are given below:

1. Enzyme + Lactose →Enzyme-Lactose complex
2. Enzyme-Lactose →Galactosyl-Enzyme + Glucose
3. Galactosyl-Enzyme + H_2O →Galactose + Enzyme

or

Galactosyl-Enzyme + Acceptor sugar →Oligosaccharide + Enzyme

Table 5.2 Kinetic studies for free (soluble) and immobilized enzyme

Proposed kinetic model	Source	Enzyme form	Substrate	pH	Temperature (°C)	[S]	[E]	K_M (M)	K_I (M)	References
Michaelis–Menten first order	*Kluyveromyces fragilis*	Free	Phosphate buffer with similar salt composition of milk	6.5	45	35 mg/mL	7 mg/mL	–	–	Santos et al. (1996)
First order kinetics model and a single-step non-first-order enzyme inactivation kinetic model	*Kluyveromyces marxianus lactis*	Free	Skimmed milk	6.7	37	250 mL Skimmed milk	1.25 mL/L	–	–	Akgue et al. (2012)
Michaelis–Menten without inhibition by product (galactose) (with enzyme–lactose complex formation)	*E. coli*	Free	Phosphate buffer pH 7.3	7.3	35	23–58 mM	0.019 mg/mL	–	–	Berrueta and García (1988)

Michaelis–Menten with competitive inhibition by product (total galactose)	*Kluyveromyces fragilis*	Free	Phosphate buffer with similar salt composition of milk	6.5	40	35 mg/mL	7 mg/mL	0.023	0.0153	Santos et al. (1998)
	E.coli	Microencapsulated enzyme	Phosphate buffer, pH 7.3	7.3	25–37	0.5–3 mM	0.0308 mg/mL	–	–	Wadiak and Carbonell (1975)
	Aspergillus oryzae	Immobilized	–	6.5	50	–	–	0.160	0.0045	Friend and Shahani (1982)
	Aspergillus oryzae	Free	Milk	6.5	30	50 mM	5 mg/mL	0.105–0.150	0.005–0.030	Portaccio et al. (1998)
	Kluyveromyces fragilis	Free	Phosphate buffer	6.86	43	146–292 mM	0.021–0.088 mg/mL	0.0436	0.0519	Carrara and Rubiolo (1996)
		Immobilized	Phosphate buffer	6.86	43	146–292 mM	0.021–0.088 mg/mL	0.1370	0.2340	Carrara and Rubiolo (1996)
	Kluyveromyces lactis	Free	Phosphate buffer	7.1	25	0.1–3.0 mg/mL	12 mg/mL	0.00254	0.01088	Fontes et al. (2001)
	Aspergillus niger	Immobilized	Citrate buffer	3.6	40	14.4–144 mM	4.2–472 UI	0.0189	0.000468	Papayannakos et al. (1993)
	Aspergillus niger	Free	Lactic acid Buffer pH 4.0	4.0	8	73–584 mM	10–10,000 (mg/L)	0.0833	0.000470	Yang and Okos (1989)
	Aspergillus niger	Free and immobilized	Lactose solution	3.6	55	0–110 g/L	1–15 mg/L	0.050	0.0015–0.003	Mirsalami and Alihosseini (2021)

(continued)

Table 5.2 (continued)

Proposed kinetic model	Source	Enzyme form	Substrate	pH	Temperature (°C)	[S]	[E]	K_M (M)	K_I (M)	References
Michaelis–Menten with competitive inhibition by product (α- and β- galactose)	*Aspergillus niger*	Free	Acidic whey	–	50	25–500 mM	250–5000 (mg/L)	–	–	Flaschel et al. (1982)
	–	Immobilized	Lactose solution and 0.5 g sodium acetate/L	4.5	40	–	50–150 g/L	–	–	Peterson et al. (1989)
	B. circulans	Immobilized	Skimmed milk	–	20–50	146 mM	1.7 mg/cm^3 membrane	–	–	Bakken et al. (1989)
Michaelis–Menten with competitive inhibition by product (glucose)	*Kluyveromyces lactis*	Free	–	–	–	–	–	–	–	Cavaille and Combes (1995)
Empirical model for glucose-formation kinetics	*Bifidobacterium bifidum*	Free	Milk	4.5	45–50	–	–	–	–	Voget et al. (2022)
Monte Carlo method	*Beta galactosidase*	Immobilized	Lactose of 100 g/L	5.0	55	100 g/L	12 U/mL	–	–	Gao et al. 2019
Di-, oligosaccharide formation	*Aspergillus oryzae*	Free	Lactose solution	4.5	40	0·139 to 1·67 mol/L	4.2 U/mL	–	–	Iwasaki et al. (1996)

Table 5.3 Kinetic equations for fitting the experimental data for lactose hydrolysis

Kinetic model	Kinetic equation	References
First order	$v = k\,[E][S]$	Ladero et al. (2000)
	$\frac{d(CL)}{dt} = -k[CL], CL = CLo * ex(-k * t)$	Akgue et al. (2012)
Michaelis–Menten	$v = \frac{(Vmax * [S])}{Km + [S]}$	Mirsalami and Alihosseini (2021)
Michaelis–Menten with two parameters	$v = \frac{k[E][S]}{Km + [S]}$	Ladero et al. (2000)
Michaelis–Menten with competitive inhibition by lactose	$v = \frac{k[E][S]}{Km + [S]\left(1 + \frac{[S]}{Ki}\right)}$	Ladero et al. (2000) and Mirsalami and Alihosseini (2021)
Michaelis–Menten with acompetitive inhibition by product (galactose)	$v = \frac{k[E][S]}{Km\left(1 + \frac{[P]}{Ki}\right) + [S]}$	Ladero et al. (2000)
	$v = \frac{(Vmax * [S])}{Km\left(1 + \frac{[P]}{Ki}\right) + [S]}$	Wadiak and Carbonell (1975)
	$\frac{d(L)}{dt} = \frac{Vmax\,L}{L + Km\left(1 + \frac{[G]}{Ki}\right)}$	Fontes et al. (2001)

(continued)

Table 5.3 (continued)

Kinetic model	Kinetic equation	References
Michaelis–Menten with acompetitive inhibition by produce	$v = \frac{k[E][S]}{Km + [S]\left(1 + \frac{[P]}{Ki}\right)}$	Ladero et al. (2000)
Michaelis–Menten with no competitive inhibition by product	$v = \frac{k[E][S]}{\left(Km + [S])(1 + \frac{[P]}{KI}\right)}$	Ladero et al. (2000)
Michaelis–Menten with mixed inhibition by product	$v = \frac{k[E][S]}{Km\left(1 + \frac{[P]}{Ki}\right) + [S]\left(1 + \frac{[P]}{K'i}\right)}$	Ladero et al. (2000)

V = Reaction rate: [E] = Enzyme concentration, [S] = Substrate concentration, Km = Michaelis–Menten constant
K_i = the inhibition constant, [P] = Concentration of product, L = Concentration of Lactose, G = Concentration of galactose
[CL] = Residual lactose concentration (g/ L), [CL0] = Lactose concentration (g / L) at t = 0, k = Hydrolysis kinetic constant (min^{-1})

During lactose hydrolysis, GOS are produced by transgalactosylation of β-galactosidase, a process in which β-galactosidase transfers galactose moiety to another sugar, likely being lactose instead of transferring the galactose moiety to the hydroxyl functional group of H_2O molecule. Lactose and other sugars are serving as galactosyl acceptors when it causes galacto oligosacchaide production with different level of polymerization and glycosidic bonds (Torres et al., 2010). Parameters such as pH, temperature, concentration of lactose and source of *β*-galactosidase play an important role in determining amount and type of GOS synthesis. The detailed hydrolytic and transgalactosylation mechanism catalysed by β-galactosidase enzyme is explained in Chap. 3. β-galactosidase obtained from different sources may have different properties for producing glycosidic linkages and resulting GOS production with varying degrees of polymerization.

Boon et al., 2000 quantified the concentration of oligosaccharides (OS) produced with variation of temperature and enzyme origin. β-galactosidase of *Bacillus circulans* was found to be most abundant, most different and largest size of OGs whereas that of *Kluveromyces* spp. were mainly produces trisaccharides. Bosso et al. (2016) investigated the hydrolysis efficiency of *K. lactis* and *A. oryzae* in lactose substrate, UHT milk and skimmed milk at different concentration of (0.7, 1.0 and 1.5%) and temperature of (30, 35, 40 & 45 °C). It was observed that *K. lactis* has higher hydrolysis rate as compared to *A. oryzae*. Use of β- galactosidase from the source *K. latcis*, glycosidic linkages were found most common between galactose β1 → 6 glucose bond. Pal et al. (2013) extracted, purified and characterized β-galactosidase from almond thermodynamically at optimum pH 5.5 with lactose hydrolysis of ~50% in milk. Shi et al., 2022 analyzed the activity of *β*-galactosidase from *A. oryzae* at acidic pH optima and observed severe product inhibition which limit enzyme activity in lactose hydrolysis in dairy products. The researchers conducted "structure-based sequence alignment" so as to mutate to shift pH optima of lacA from 4.5 to 5.5 and 6.0, respectively. This study provided a novel strategy for using genetic mutation as an engineering technique for manipulating pH optima of glycoside hydrolases. Zhong et al. (2023) isolated *β*-galactosidase from lotus seeds at optimum pH 4.0 and optimum temperature 50 °C. The study showed higher lactose hydrolysis and transgalactosylation properties when lotus seed powder is added to milk or whey and also the *β*-galactosidase was not inhibited by the higher concentration of galactose.

The choice of enzyme source for GOS production from whey has been found to significantly influence the yield and composition of GOS. Several studies have investigated the impact of different microbial sources of enzymes on GOS synthesis from whey. In the study conducted by Goulas et al., (2007), lactase derived from the microbial source *Bifidobacterium bifidum* was utilized to catalyze the hydrolysis of lactose present in concentrated sweet whey permeate, which initially contained 45% lactose. The researchers observed a notable disparity in the distribution of GOS with different degrees of polymerization (DP) when comparing the concentrated sweet whey permeate to a lactose buffer solution. Specifically, the microbial lactase exhibited a higher proportion of DP2 (48%) and a lower proportion of DP3 (52%) in the whey-concentrated sweet whey permeate, while the lactose buffer solution yielded a composition of 43% DP2 and 57% DP3 GOS. These findings suggest that the presence of whey concentrates exerts an influence on the enzymatic hydrolysis process, leading to distinct proportions of DP2 and DP3 GOS. Fischer and Kleinschmidt (2018) investigated the variation in GOS in different substrates, including buffer, acid, and sweet whey (20% concentration). They used the fungus *Aspergillus oryzae* to hydrolyze the whey substrates. The results showed reduced variability in DP2 GOS in all three substrates, indicating a consistent production of DP2 GOS across the different whey types. However, a higher percentage of GOS disaccharides was formed in sweet whey compared to the other substrates.

In the study by Mano et al. (2019), *Kluyveromyces lactis* was evaluated alongside *Aspergillus oryzae* and *Escherichia coli* strain K12 for galactooligosaccharides (GOS) production using whey permeate as a substrate. *Kluyveromyces lactis* exhibited superior efficiency compared to alternative enzyme sources, yielding approximately 25 g of GOS per 100 g of lactose. Conversely, *E. coli* strain K12 demonstrated markedly lower GOS production, suggesting diminished transgalactosylation activity in this strain.

In 2021, Rico-Rodriguez et al. investigated GOS production utilizing two distinct sources of β-galactosidase (*A. oryzae* and *K. lactis*) either individually or in a 1:1 ratio through simultaneous enzymatic activity. This investigation was conducted in cheese whey, as well as permeate and lactose solutions. The findings demonstrated an elevated GOS yield compared to that achieved with individual enzyme sources. Subsequently, in a separate study conducted by the same research group, cheese whey, whey permeate, and lactose solutions were evaluated for GOS production using a similar enzymatic approach, either alone or in combination. Interestingly, GOS production employing a combination of enzymes was comparable to that achieved using individual enzymes. Once again, β-galactosidase sourced from *Kluyveromyces* species exhibited heightened effectiveness in GOS production, particularly evident in cheese whey substrates. Liburdi and Esti (2022) found that β-galactosidase obtained from thermophilic microorganism is more suitable for transgalactosylation process at relatively high temperature, reason being the inhibition of microbial contamination.

It was concluded that initial lactose concentration has a direct relationship to the amount of GOS produced. By increasing the temperature of the reaction in a sub-saturated lactose concentration, the GOS production increases. Hence, in order to optimize GOS production, at lower water content, higher lactose concentration, and the higher temperature of reaction typically yields a higher of GOS production.

5.4 Effect of Process Parameters on Lactose Hydrolysis

5.4.1 Effect of pH

The pH of the reaction environment has a significant effect on the lactose hydrolysis process and the resulting end products. The influence of pH on enzyme stability and activity is widely recognized, as the majority of enzymes exhibit a strong pH dependence, wherein the pH of the solution plays a crucial role in determining their stability and functional efficiency too. β-galactosidase exhibits optimal activity within a specific pH range. Generally, lactase activity is highest at a slightly acidic to neutral pH, typically around pH 6.0 to 6.5. Within this range, the enzyme is most efficient in catalysing the hydrolysis of lactose into glucose and galactose. Deviations from this pH range can reduce the enzyme's activity, leading to slower lactose hydrolysis. From various research studies it is revealed that, the enzyme obtained from bacterial sources shows their relative activity for lactose hydrolysis and GOS production at a pH range 6.5–7, however enzyme obtained from yeast and fungi sources shows their optimal activity at pH range 4.19–7.0 (Deshmukh et al., 2024).

Furlan et al. (2000) have compared properties of free and immobilized β-galactosidase (*A. oryzae*), entrapped in alginate–gelatin beads and cross-linked with glutaraldehyde. *A. oryzae* β-galactosidase enzyme (both the free and immobilized forms of the enzyme) exhibited complete retention of their activity even after 18 h of incubation, within a pH range spanning from 4.5 to 7.0 The enzyme demonstrated its highest enzymatic activity within a temperature range of 45 to 60 °C, while the optimal pH values for maximum activity fell between 4.0 and 5.0. The pH value that corresponded to the highest activity of the free β-galactosidase enzyme was 4.5, whereas the immobilized form of the enzyme displayed optimal activity at a pH of 5.0. It was also concluded that free and immobilized enzymes shown same profiles but the optimal temperature of immobilized enzyme was higher compare to free enzyme.

The pH of the reaction environment can influence the distribution of hydrolysis products. At higher pH values, the formation of glucose is favoured, resulting in a higher glucose-to-galactose ratio. On the other hand, at lower pH values, the formation of galactose is favoured, leading to a higher galactose-to-glucose ratio. The pH of the reaction environment also affects the stability of the lactase enzyme. Enzyme stability is essential for maintaining its activity over the course of the reaction. Extreme pH values, either highly acidic or highly alkaline, can denature the enzyme, leading to a loss of activity. Therefore, it is crucial to operate within the pH range that maintains the stability and activity of lactase. The solubility of lactose can be a pH-dependent. At certain pH values, lactose solubility may be enhanced or reduced. Changes in lactose solubility can affect the accessibility of the enzyme to the substrate and consequently influence the rate and extent of lactose hydrolysis. Popescu et al. (2021) studied the effect of pH on lactose hydrolysis tendency of *L.lactis* and *B. licheniformis* and observed that at a pH of 6.5 the lactose hydrolysis degree for the enzyme from former was maximum (77.6%), the decrease of the pH below the value of 6.5 leads to the inactivation of the enzyme, demonstrating sensitivity to low values of pH. For the enzyme from *B. licheniformis* the maximum degree of lactose hydrolysis (85.2%) was reached at pH values of 5.5–6.0, the decrease of the pH below the value of 5.5 leads to inactivation of the enzyme. Study conducted by Singh et al. (2022), enzyme derived from *K. lactis* was found more effective than the *B. lichenformis* for lactose hydrolysis and GOS formation in 16% lactose solution at pH 6.7 at 7 °C. *K. lactis* exhibited 96.61% lactose hydrolysis and 7.28% GOS production after 12 h of reaction in concentrated skim milk with 40% total solids.

Study conducted by Rico-Rodríguez et al. (2020), obtained a highest yield of GOS production (32–46%) using enzyme from *K. lactis* source with 40% lactose solution at 42 °C and pH 6.0. Jenab et al. (2018) reported 23% GOS formation in 40% lactose solution with 0.05% β-galactosidase from *A. oryzae* incubated at 45 °C and pH 4.5.

5.4.2 Effect of Temperature

The rate of lactose hydrolysis increases with an increase in temperature. This is because of higher temperatures provide more energy to the enzymes responsible for lactose hydrolysis, allowing them to work more efficiently. Consequently, higher temperatures lead to faster lactose conversion. Like most enzymes, β-galactosidase has an optimal temperature range in which it functions most effectively. For β-galactosidase, this optimal range is usually around human body temperature, which is approximately 37 °C. At this temperature, β-galactosidase has its highest activity, leading to the most rapid lactose hydrolysis. While higher temperatures initially increase the reaction rate, excessively high temperatures can lead to the denaturation of lactase. Denaturation is the process by which the enzyme's structure is disrupted, causing it to lose its activity. If β-galactosidase denatures due to excessively high temperatures, lactose hydrolysis will slow down or cease altogether. Temperature also affects the shelf life and storage of lactose-hydrolyzed dairy products. When stored at lower temperatures, the hydrolysis process continues at a slower rate, leading to a longer shelf life. On the other hand, if stored at higher temperatures, the hydrolysis process will proceed more rapidly, potentially reducing the product's shelf life. Yin et al., (2017) suggested that the initial substrate concentration and the reaction temperature for transgalactosylation activity by the enzyme has great impact on the final yield of GOS production in the lactose containing solution or milk and milk products. Higher temperature can be used for while producing GOS production in buffer solution but the same temperature cannot be treated for milk and milk products, as it may promote the microbial growth and shelf life may affect. In the food industry, manufacturers often control the temperature during lactose hydrolysis to achieve the desired level of lactose conversion while ensuring the product's overall quality and shelf life. Proper temperature management ensures that lactose-hydrolyzed dairy products are safe, easily digestible, and maintain their desired characteristics.

5.4.3 Effect of Substrate Concentration

The substrate concentration, specifically the initial lactose concentration in the reaction mixture, can have an effect on lactose hydrolysis and the resulting product composition. The effect of substrate concentration on lactose hydrolysis products can be summarized as follows (Dantas et al., 2021; Singh et al., 2022):

(a) **Reaction rate**: Generally, increasing the lactose concentration in the reaction mixture leads to a higher reaction rate. This is because a higher substrate concentration provides more lactose molecules for the enzyme to act upon, increasing the frequency of enzyme-substrate collisions and facilitating a faster hydrolysis process. As a result, the production of glucose and galactose is accelerated.
(b) **Extent of hydrolysis**: The substrate concentration can also influence the extent of lactose hydrolysis. Higher lactose concentrations may require longer reaction times to achieve complete hydrolysis, especially if the enzyme becomes saturated. Therefore, it is important to balance the lactose concentration with the reaction time to achieve the desired level of hydrolysis and product composition.
(c) **Product distribution**: The lactose concentration can affect the distribution of hydrolysis products. At lower lactose concentrations, the production of glucose and galactose occurs whereas at higher lactose concentration, transgalactosylation reaction prevails which leads to the formation of GOS. However, there may be limits beyond which further increases in lactose concentration do not result in proportional increases in glucose and galactose formation. This can be due to enzyme saturation or inhibition effects at extremely high lactose concentrations.
(d) **Enzyme Inhibition**: In some cases, excessively high lactose concentrations can inhibit the activity of the β-galactosidase enzyme. This inhibition can occur due to product inhibition, where the accumulation of glucose and galactose inhibits the enzyme's activity. Therefore, there may be an optimal lactose concentration range where the enzyme activity is not inhibited, and efficient lactose hydrolysis can occur.

Optimizing the substrate concentration in lactose hydrolysis is essential to achieve the desired product composition and reaction efficiency. It involves balancing the lactose concentration to ensure an adequate reaction rate without encountering enzyme saturation or inhibition effects. Determining the appropriate lactose concentration often requires experimental investigation and consideration of factors such as enzyme concentration, reaction time, and the specific lactase enzyme used.

5.5 Effect of Glucose and Galactose Concentration

The presence of glucose can affect the rate of lactose hydrolysis in the following ways (Dantas et al., 2021; Singh et al., 2022):

(a) **Inhibition of lactase activity**: High concentrations of glucose can inhibit the activity of lactase. Glucose can bind to the enzyme at active site and prevent lactose from binding, reducing the rate of hydrolysis. This inhibition is known as competitive inhibition, where glucose and lactose compete for the same binding site on the enzyme.

(b) **Induction of glucose transporters**: Glucose can induce the expression of glucose transporters on the cell membrane. These transporters facilitate the entry of glucose into the cell. When glucose concentration is high, the cells may prioritize the uptake of glucose over lactose. This can reduce the availability of lactose to lactase, leading to a decrease in lactose hydrolysis rate.
(c) **Altered osmolarity**: High concentrations of glucose can increase the osmolarity of the solution. Changes in osmolarity can affect the activity of enzymes and the stability of the surrounding environment. Extreme osmolarity conditions may lead to changes in the enzyme's conformation, affecting its catalytic activity.

When galactose concentration increases, it can affect lactose hydrolysis in the following ways:

(a) **Competitive inhibition**: Galactose can competitively inhibit the action of lactase enzyme by binding to its active site. This can reduce the rate of lactose hydrolysis as the lactase enzyme has a higher affinity for galactose than lactose. As a result, higher galactose concentrations can slow down the hydrolysis of lactose.
(b) **Saturation of enzyme**: If the concentration of galactose is extremely high, it can saturate the available β-galactosidase enzyme molecules. When the enzyme is saturated, increasing the concentration of galactose further will not increase the rate of lactose hydrolysis. This phenomenon is known as substrate saturation.
(c) **Reverse reaction**: In some cases, high concentrations of galactose can lead to the reverse reaction, where galactose and glucose molecules combine to form lactose. This occurs when the concentration of galactose is significantly higher than glucose. In this situation, the net hydrolysis of lactose may be reduced or even reversed, resulting in the formation of lactose instead of its breakdown.

5.6 Effect of Mineral and Ion Concentration

The effect of cations such as potassium and sodium ions also affect the enzyme activity in concentrations normally observed in a natural substrate. Garman et al. (1996) found observed that Na^+ and K^+, Mg^{+2} contributing the maximum rates of lactose hydrolysis by ß-galactosidase obtained from the source *B. bifidum* in buffer solutions. Voget et al. (2022), studied the effect of different cations on recombinant lactase from *Bifidobacterium bifidum* (RBBL) activity in which the lactose hydrolysis was analysed in 5.1% (w/v) lactose in phosphate solution (PPB) (containing 50 mM K^+) at pH 6.6 and milk with or without the addition of different cations (Ca^{+2}, Mg^{+2}, Mn^{+2},Na^+). These cations were selected and analysed according to their occurence in milk (Fuente & Juárez, 2015). The yeast neutral lactase (YNL) was also as a reference in the experiments. Study revealed that, the addition of cations has shown no significant effect ($p < 0.05$) on RBBL activity, either in the buffer solution or in milk. The results with YNL are also reliable with the reported properties of enzyme. All the cations were affected that of enzyme activity in some way. In PPB, the addition of Mg^{+2} and Mn^{+2} was found in increased lactose hydrolysis, while Na^{+} has shown inhibitory effect. A similar response was observed in milk where the enzyme activity increased with the addition of Mn^{+2} (18.5%), Mg^{+2} (22%) and K^{+} (15%) and was decreased with addition of Na^{+} (−26%) and Ca^{+2} (−15%). The increase in activity that was observed when K^{+} was added to milk which can be explained that the enzyme activity was depends on the concentration ratio of these cations instead of the absolute concentration of Na^{+} (Flores et al., 1996). In another study conducted by Fischer and Kleinschmidt (2018), the impact of cations on the enzyme activity of β-galactosidase using two different combinations of either *Aspergillus oryzae* and *Cryptococcus laurentii* in presence of substrate *o*-nitrophenyl- β -D-galactopyranoside (oNPG). Their study observed that the activity of enzyme was notably influenced by the presence of different cations in the surrounding environment. Specifically, they observed that calcium had no discernible effect on the enzyme's activity, while magnesium exhibited a moderate impact. In contrast, potassium, sodium, and ammonium were identified as strong activators of the enzyme. Regarding Ao, the researchers noted that cations had a minimal effect on its activity; however, it exhibited a higher preference for utilizing GOS as a substrate in whey as compared to lactose solution.

5.7 Effect of Membrane Processing

Membrane applications play a significant role in various industrial processes, including enzymatic lactose hydrolysis. Because, the higher transgalactosylation activity of the enzyme *β*-galactosidase in the permeate than to milk (Chen et al., 2002). It is well reported that the greater the initial lactose concentration in the mixture the greater the conversion of oligosaccharides, which is useful for production of high quality of low-lactose containing milk and dairy products. The use of membrane processing, separates the milk proteins which limits the interference of milk proteins with *β*-galactosidase enzyme activity. Manufacturing of lactose free/low lactose milk products using membrane technology followed by conversion of residual lactose to its monosaccharide units using enzymatic hydrolysis may help to manufacturers to retain the taste and sweetness of original milk, but this technology is expensive. Actually, removing lactose from milk by applying membrane technology rises production cost by about 20–30% (Shakeel-Ur-Rehman, 2009). Enzymatic membrane bioreactors may be used for GOS production through either continuous immobilized enzyme hydrolysis or batch hydrolysis with the use of membrane filtration such as ultrafiltration, microfiltration, nanofiltration and reverse osmosis.

Holst and Lauritzen (2013) (US patent USOO8445052B2), invented a method for development substantially lactose-free milk product (less than 0.01% lactose), which was produced from original milk with retaining the taste and sweetness of the original milk. In this invention, milk components were separated using ultrafiltration, nanofiltration, and reverse osmosis followed by hydrolysis of lactose-free milk (UF retentate) with the optional added water or whey salts. Chen et al. (2002), were optimized the enzymatic process for manufacturing low lactose milk containing higher yield of total oligosaccharides was 31.1% using UF permeate, while in milk it was 22.8%. The optimal conditions for the enzymatic reaction in the permeate was 25.3% of initial lactose concentration, an E/S ratio of 6.7%, at 50 °C for 3.5 h. Study also revealed that, *β*-galactosidase possess more activity in the permeate than to milk, as milk proteins were removed by the ultrafiltration membrane which could be inferred that milk proteins might inhibit the *β*-galactosidase activity. Jakubowski et al. (1975) suggested that serum (whey) proteins may decrease the *β*-galactosidase activity obtained from *A. niger*. Mozaffar et al. (1985) also reported that the milk proteins, decreased *β*-galactosidase activity obtained from the source *E. coli* and *K. lactis. However, the inhibition mechanism by milk proteins is still unknown. Second, possible basis for the improved β-galactosidase* activity in the permeate than to milk was the differences in the mineral composition among these two solutions. There has been much research focused on the effects of Mg^{2+}, Ca^{2+}, Na^{2+} on the *β*-galactosidase activity from the various sources with reported contradictory reports (Hustad et al., 1973; Agbebavi et al., 1987). Agbebavi et al. (1987) reported that metal ions such as Mg^{2+} *and* Ca^{2+}, activated *β*-galactosidase activity. Itoh et al. (1982) observed that Mg^{2+}, Ca^{2+}, and Zn^{2+} inhibited *β*-galactosidase activity. These contradictory results regarding the effects of mineral ions could be due to the variation in the enzyme sources or experimental conditions. Mozaffar et al. (1985) stated that Ca^{2+} and Na^{2+} decreased *β*-galactosidase activity obtained from the sources E. coli and *K. lactis*, but not from *B. circulans*. Hence, the membrane processing such as UF and RO for the production of low-lactose milk containing higher oligosaccharides could also be preventing the protein damage and enhances the flavour qualities due to the lack of heating for obtaining the concentrated milk and restricts the direct contact between enzyme and milk proteins. Córdova et al. (2016) used UF-Membrane bioreactor for enzymatic hydrolysis using free β-galactosidase from *A. niger* in high concentrations of lactose (470 g/L) for synthesis of GOS. Study revealed that, 2.44-fold higher GOS synthesis observed than batch hydrolysis of lactose at the optimum conditions of using UF process at transmembrane-pressure (4.38 bar), crossflow-velocity (7.35 m/s), and temperature (53.1 °C). Kaczyński et al. (2019), produced a value-added fermented product by goat milk and its MF/UF permeate with β-galactosidase obtained from *Kluyveromyces lactis*. The highest GOS content was achieved at a temperature 37 °C at 20 min at 4.4 pH. In another study conducted by Aquino et al. (2017), the recovery of a high purity caprine oligosaccharides (COS) concentrate was obtained by complete lactose hydrolysis (with food grade fungal lactase derived from *Aspergillus oryzae*), fermentation of the resulting monosaccharides and concentration by nanofiltration. The sample of defatted goat whey was first ultrafiltered with discontinuous diafiltrations inorder to enhance the recovery of COS in whey permeate followed by concentration by application of nanofiltration. COS recovery was achieved about of 75% with negligible amounts of monosaccharides (0.3% of the initial amount of lactose in the whey permeate). Final retentate contains 27 g/L of oligosaccharides of which 67.6 and 34.4% of acidic and neutral oligosaccharides respectively which were obtained from caprine milk.

5.8 Effect of Ultrasonication

Ultrasonication is the application of high-frequency sound waves to a liquid or solution (Gajendragadkar & Gogate, 2017). Usually, ultrasound waves are kind of sound waves with a frequency above the hearing limit of human ear ($\approx$20 kHz). Now a day, the use of ultrsounication process aroused researchers concern in order to study the influence on the lactase enzyme activities. Ultrasound has effect on inhibiting or accelerating the effects on enzymes-substrates interactions between the enzymes and substrates. Under the certain conditions of low intensity and suitable frequency, ultrasound generates cavitation, magnetostrictive effect and mechanical oscillation effect which may improve the vibrational momentum of the enzymes and improves the contact of substrate with the enzyme active cite. As a result, the biological activity of enzymes is enhanced (Wang et al., 2015). Combining the lactose hydrolysis process with ultrasonication can have several potential benefits (Demirhan & Özbek, 2009; Gajendragadkar & Gogate, 2017; Novalin et al., 2005; Şener et al., 2006) in below ways:

(a) **Enhanced reaction rate:** Ultrasonication can accelerate the rate of lactose hydrolysis by providing mechanical agitation and disrupting the mass transfer limitation. The high-frequency sound waves create cavitation bubbles in the solution, leading to the formation of micro jets and shockwaves, which can increase the contact between the lactose and the hydrolysing enzyme or catalyst.
(b) **Improved efficiency:** The combination of lactose hydrolysis and ultrasonication can increase the efficiency of the hydrolysis process. The enhanced mass transfer and increased surface area resulting from ultrasonication can lead to better access of the enzyme or catalyst to the lactose molecules, resulting in more effective hydrolysis.
(c) **Reduced reaction time:** Ultrasonication can significantly reduce the reaction time required for lactose hydrolysis. The mechanical and chemical effects induced by ultrasound waves can promote the breakage of glycosidic bonds in lactose, allowing for faster conversion into glucose and galactose.
(d) **Better product quality:** Ultrasonication can improve the sensory properties of the hydrolysed lactose products. The application of ultrasound can modify the structure and properties of molecules, resulting in improved taste, texture, and solubility of the hydrolysed lactose.

It's important to note that the specific parameters for combining lactose hydrolysis with ultrasonication, such as power level, duration, and enzyme or catalyst concentration, need to be optimized for each particular application. Additionally, the equipment used for ultrasonication should be compatible with food-grade materials to ensure the safety and quality of the final product (Gajendragadkar & Gogate, 2017). It has been reported free enzymes activity increased when conditions of mild ultrasound and irradiation were applied (Novalin et al., 2005, Şener et al., 2006). Rico-Rodríguez et al. (2020), investigated the effect of ultrasound-assisted GOS and gluconic acid production using different lactose source such as pure lactose, whey permeate and cheese whey. In study, it was observed that ultrasonication does not change the β–galactosidase enzyme activity obtained from the source *Kluyveromyces lactis* but the GOS yield has been increased from 35–40% when cheese whey was used as a substrate as compared to other lactose sources. Recently, Córdova et al. (2023), evaluated the effect of different ultrasonic conditions on th enzymatic production of GOS using native whey as a substrate obtained using casein micelles microfiltration by using two commercial enzyme sources as *Aspergillus oryzae* and *Kluyveromyces var lactis*. It was observed that, above 100 g/L, solute concentration significantly decreased the ultrasonic intensities, but the ultrasound wave attenuation can be reduced if the percentage of wave amplitude is increased at 70% and 0.6 s/s duty cycle. At 40 °C, the enzyme from the source *Aspergillus oryzae* showed greater activity as compared to the control when the ultrasonic intensity was below 11 W/cm^2. In most cases, these improvements lasted a few hours, then enzymes suffered faster inactivation. Meanwhile, at 30 °C, the enzyme from Kluyveromyces var. lactis was prone to enzyme inactivation, even though a low UI was applied, which may be due to its homodimeric 3D structure, resulting in inadequate results for the pursued objectives.

5.9 Conclusion

The technologies for lactose hydrolysis have been advanced significantly, however, there are still some drawbacks associated with present technologies. Present technologies for lactose hydrolysis can suffer from enzyme inhibition due to factors like high temperatures, low pH, or the presence of certain substances in the reaction mixture. Enzyme inhibition can reduce the efficiency and effectiveness of lactose hydrolysis, leading to lower conversion rates and longer reaction times. Enzymes used for lactose hydrolysis are expensive, and their availability is limited. The cost of enzymes contributes to the overall production cost of lactose-free products, making them relatively more expensive compared to conventional dairy products. Some present technologies for lactose hydrolysis, such as batch reactors, may have limitations in terms of process efficiency and reaction time. High lactose concentrations in the substrate can negatively affect the efficiency of lactose hydrolysis as the reaction equilibrium can be shifted, leading to reduced lactose conversion rates. Lactose hydrolysis can affect the flavour and texture of dairy products. Disposal of these waste streams can have environmental implications, especially if not properly managed or treated. Lactose hydrolysis is an important process in the food industry, particularly for the production of lactose-free or reduced-lactose dairy products. The development of novel enzymes or improved enzyme variants with enhanced lactose hydrolysis activity, stability, and specificity can significantly impact the efficiency and economics of lactose hydrolysis processes. Overall, the future of lactose hydrolysis technologies lies in the development of advanced enzymes, innovative reactor designs, sustainable approaches, and process optimization strategies. These advancements have the potential to improve the efficiency, cost-effectiveness, and sustainability of lactose hydrolysis, leading to a wider range of lactose-free products and enhanced nutritional options for consumers.

References

Agbebavi, J. T., Rouleau, D., & Mayer, R. C. (1987). The effects of whey cations on the kinetic parameters of a model for the enzymic hydrolysis of lactose. *Journal of Chemical Technology & Biotechnology, 37*(3), 153–168.

Akgue, L. F. B., Demirhan, E., & OeZBEK, B. (2012). A modelling study on skimmed milk lactose hydrolysis and β-galactosidase stability using three reactor types. *International Journal of Dairy Technology, 65*(2), 217–231.

Aliyah, A., Alamsyah, G., Ramadhani, R., & Hermansyah, H. (2017). Production of α-amylase and β-glucosidase from aspergillus Niger by solid state fermentation method on biomass waste substrates from rice husk, bagasse and corn cob. *Energy Procedia, 136*, 418–423.

Ansari, S. A., & Husain, Q. (2010). Lactose hydrolysis by β galactosidase immobilized on concanavalin A-cellulose in batch and continuous mode. *Journal of Molecular Catalysis B: Enzymatic, 63*(1–2), 68–74.

Aquino, L. F., de Moura Bell, J. M. L. N., Cohen, J. L., Liu, Y., Lee, H., de Melo Silva, V. L., Domizio, P., Juniour, C. A. C., & Barile, D. (2017). Purification of caprine oligosaccharides at pilot-scale. *Journal of Food Engineering, 214*, 226–235.

Bakken, A. P., Hill, C. G., Jr., & Amundson, C. H. (1989). Hydrolysis of lactose in skim milk by immobilized β-galactosidase in a spiral flow reactor. *Biotechnology and Bioengineering, 33*(10), 1249–1257.

Berrueta, J., & García, T. (1988). Hidro'lisis enzima'tica de lactosa en reactores de lecho fijo. *Ing. Quı'mica, 236*, 141–146.

Boon, M. A., Janssen, A. E. M., & Van't Riet, K. (2000). Effect of temperature and enzyme origin on the enzymatic synthesis of oligosaccharides. *Enzyme and Microbial Technology, 26*(2–4), 271–281.

Bosso, A., Morioka, L. R. I., Santos, L. F. D., & Suguimoto, H. H. (2016). Lactose hydrolysis potential and thermal stability of commercial β-galactosidase in UHT and skimmed milk. *Food Science and Technology, 36*, 159–165.

Carrara, C. R., & Rubiolo, A. C. (1996). Determination of kinetics parameters for free and immobilized β-galactosidase. *Process Biochemistry, 31*(3), 243–248.

Cavaille, D., & Combes, D. (1995). Characterization of β-galactosidase from Kluyveromyces lactis. *Biotechnology and Applied Biochemistry, 22*(1), 55–64.

Chen, C. S., Hsu, C. K., & Chiang, B. H. (2002). Optimization of the enzymic process for manufacturing low-lactose milk containing oligosaccharides. *Process Biochemistry, 38*(5), 801–808.

Córdova, A., Astudillo, C., Vera, C., Guerrero, C., & Illanes, A. (2016). Performance of an ultrafiltration membrane bioreactor (UF-MBR) as a processing strategy for the synthesis of galacto-oligosaccharides at high substrate concentrations. *Journal of Biotechnology, 223*, 26–35.

Córdova, A., Astudillo-Castro, C., Henriquez, P., Manriquez, N., Nuñez, H., Guerrero, C., et al. (2023). Ultrasound-assisted enzymatic synthesis of galacto-oligosaccharides using native whey with two commercial β-galactosidases: Aspergillus oryzae and Kluyveromyces var lactis. *Food Chemistry, 426*, 136526.

Dantas, A., Quinteros, G. J., Darvishvand, S. Y., Blanco, M., Hernandez, E., Prudencio, E. S., & Samsuri, S. (2021). The combined use of progressive and block freeze concentration in lactose-free milk: Effect of process parameters and influence on the content of carbohydrates and proteins. *Journal of Food Process Engineering, 44*(11), e13867.

Demirhan, E., & Özbek, B. (2009). A modeling study on hydrolysis of lactose recovered from whey and β-galactosidase stability under sonic treatment. *Chemical Engineering Communications, 196*(7), 767–787.

Deshmukh, N., Rao, P. S., Sharma, H., & Kumar, M. (2024). Waste to nutrition: The evolution of whey, a byproduct to galactooligosaccharides production. *Food Chemistry Advances, 4*, 100642.

Fischer, C., & Kleinschmidt, T. (2018). Combination of two β-galactosidases during the synthesis of galactooligosaccharides may enhance yield and structural diversity. *Biochemical and Biophysical Research Communications, 506*(1), 211–215.

Flaschel, E., Raetz, E., & Renken, A. (1982). The kinetic of lactose hydrolysis for the β-galactosidase from *Aspergillus Niger. Biotechnology and Bioengineering, 24*(11), 2499–2518.

Flores, M. V., Ertola, R. J., & Voget, C. E. (1996). Effect of monovalent cations on the stability and activity of *Kluyveromyces lactis* β-galactosidase. *LWT-Food Science and Technology, 29*(5–6), 503–506.

Fontes, E. A., Passos, F. M., & Passos, F. J. (2001). A mechanistical mathematical model to predict lactose hydrolysis by β-galactosidase in a permeabilized cell mass of Kluyveromyces lactis: Validity and sensitivity analysis. *Process Biochemistry, 37*(3), 267–274.

Freitas, F. F., Marquez, L. D., Ribeiro, G. P., Brandão, G. C., Cardoso, V. L., & Ribeiro, E. J. (2011). A comparison of the kinetic properties of free and immobilized aspergillus oryzae β-galactosidase. *Biochemical Engineering Journal, 58*, 33–38.

Friend, B. A., & Shahani, K. M. (1982). Characterization and evaluation of aspergillus oryzae lactase coupled to a regenerable support. *Biotechnology and Bioengineering, 24*(2), 329–345.

Fuente, M. A., & Juárez, M. (2015). Milk and dairy products. In M. Guardia & S. Garrigues (Eds.), *Handbook of mineral elements in foods* (pp. 645–668).

Furlan, S. A., Schneider, A. L., Merkle, R., de Fátima Carvalho-Jonas, M., & Jonas, R. (2000). Formulation of a lactose-free, low-cost culture medium for the production of β-D-galactosidase by Kluyveromyces marxianus. *Biotechnology Letters, 22*(7), 589–593.

Gajendragadkar, C. N., & Gogate, P. R. (2017). Ultrasound assisted acid catalyzed lactose hydrolysis: Understanding into effect of operating parameters and scale up studies. *Ultrasonics Sonochemistry, 37*, 9–15.

Gao, L., Guo, Q., Lin, H., Pan, D., Huang, X., Lin, J., & Lin, J. (2019). Modeling of lactose enzymatic hydrolysis using Monte Carlo method. *Electronic Journal of Biotechnology, 40*, 78–83.

Garman, J., Coolbear, T., & Smart, J. (1996). The effect of cations on the hydrolysis of lactose and the transferase reactions catalysed by β-galactosidase from six strains of lactic acid bacteria. *Applied Microbiology and Biotechnology, 46*, 22–27.

Gerbault, P., Liebert, A., Itan, Y., Powell, A., Currat, M., Burger, J., Swallow, D. M., & Thomas, M. G. (2011). Evolution of lactase persistence: An example of human niche construction. *Philosophical Transactions of the Royal Society B: Biological Sciences, 366*(1566), 863–877.

Goulas, T. K., Goulas, A. K., Tzortzis, G., & Gibson, G. R. (2007). Molecular cloning and comparative analysis of four β-galactosidase genes from Bifidobacterium bifidum NCIMB41171. *Applied Microbiology and Biotechnology, 76*, 1365–1372.

Holst, H. H., & Lauritzen, K. (2013). (US patent USOO8445052B2). *Google patents*. Online Patent Search. Retrieved from https://patents.google.com/

Horner, T. W., Dunn, M. L., Eggett, D. L., & Ogden, L. V. (2011). β-Galactosidase activity of commercial lactase samples in raw and pasteurized milk at refrigerated temperatures. *Journal of Dairy Science, 94*(7), 3242–3249.

Hustad, G. O., Richardson, T., & Olson, N. F. (1973). Immobilization of β-galactosidase on an insoluble carrier with a polyisocyanate polymer. II. Kinetics and stability. *Journal of Dairy Science, 56*(9), 1118–1122.

Itoh, T., Suzuki, M., & Adachi, S. (1982). Production and characterization of β-galactosidase from lactose-fermenting yeasts. *Agricultural and Biological Chemistry, 46*(4), 899–904.

Iwasaki, K. I., Nakajima, M., & Nakao, S. I. (1996). Galacto-oligosaccharide production from lactose by an enzymic batch reaction using β-galactosidase. *Process Biochemistry, 31*(1), 69–76.

Jakubowski, J., Giacin, J. R., Kleyn, D. H., Gilbert, S. G., & Leeder, J. G. (1975). Effect of calcium, magnesium and whey proteins on the activity of beta-galactosidase (A. Niger) immobilized on collagen. 467–469.

Jenab, E., Omidghane, M., Mussone, P., Armada, D. H., Cartmell, J., & Montemagno, C. (2018). Enzymatic conversion of lactose into galacto-oligosaccharides: The effect of process parameters, kinetics, foam architecture, and product characterization. *Journal of Food Engineering, 222*, 63–72.

Jurado, E., Camacho, F., Luzon, G., & Vicaria, J. M. (2002). A new kinetic model proposed for enzymatic hydrolysis of lactose by a β-galactosidase from Kluyveromyces fragilis. *Enzyme and Microbial Technology, 31*(3), 300–309.

Kaczyński, Ł. K., Cais-Sokolińska, D., & Szwengiel, A. (2019). Kinetics of lactose hydrolysis and galactooligosaccharides formation in beverages based on goat's milk and its permeate. *Food Science and Biotechnology, 28*, 1529–1534.

Ladero, M., Santos, A., & Garcıa-Ochoa, F. (2000). Kinetic modeling of lactose hydrolysis with an immobilized β-galactosidase from Kluyveromyces fragilis. *Enzyme and Microbial technology, 27*(8), 583–592.

Liburdi, K., & Esti, M. (2022). Galacto-oligosaccharide (GOS) synthesis during enzymatic lactose-free Milk production: State of the art and emerging opportunities. *Beverages, 8*(2), 21.

Mano, M. C. R., Paulino, B. N., & Pastore, G. M. (2019). Whey permeate as the raw material in galactooligosaccharide synthesis using commercial enzymes. *Food Research International, 124*, 78–85.

Mirsalami, S. M., & Alihosseini, A. (2021). Selection of the most effective kinetic model of lactase hydrolysis by immobilized aspergillus Niger and free β-galactosidase. *Journal of Saudi Chemical Society, 25*(12), 101395.

Mozaffar, Z., Nakanishi, K., & Matsuno, R. (1985). Formation of oligosaccharides during hydrolysis of lactose in milk using β-galactosidase from Bacillus circulans. *Journal of Food Science, 50*(6), 1602–1606.

Nag, D., Kumar, V., Kumar, V., Kumar, S., & Singh, D. (2021). A new extracellular β-galactosidase producing kluyveromyces sp. PCH397 from yak milk and its applications for lactose hydrolysis and prebiotics synthesis. *Indian Journal of Microbiology, 61*(3), 391–395.

Novalin, S., Neuhaus, W., & Kulbe, K. D. (2005). A new innovative process to produce lactose-reduced skim milk. *Journal of Biotechnology, 119*(2), 212–218.

Pal, A., Lobo, M., & Khanum, F. (2013). Extraction, purification and thermodynamic characterization of almond (Amygdalus communis) β-galactosidase for the preparation of delactosed milk. *Food Technology and Biotechnology, 51*(1), 53–61.

Papayannakos, N., Markas, G., & Kekos, D. (1993). Studies on modelling and simulation of lactose hydrolysis by free and immobilized β-galactosidase from aspergillus Niger. *The Chemical Engineering Journal, 52*(1), B1–B12.

Peterson, R. S., Hill, C. G., Jr., & Amundson, C. H. (1989). Effects of temperature on the hydrolysis of lactose by immobilized β-galactosidase in a capillary bed reactor. *Biotechnology and Bioengineering, 34*(4), 429–437.

Popescu, L., Bulgaru, V., & Siminiuc, R. (2021). Effect of temperature, pH and amount of enzyme used in the lactose hydrolysis of milk. *Food and Nutrition Sciences, 12*(12), 1243–1254.

Portaccio, M., Stellato, S., Rossi, S., Bencivenga, U., Eldin, M. M., Gaeta, F. S., & Mita, D. G. (1998). Galactose competitive inhibition of β-galactosidase (Aspergillus oryzae) immobilized on chitosan and nylon supports. *Enzyme and Microbial Technology, 23*(1–2), 101–106.

Rico-Rodríguez, F., Villamiel, M., Ruiz-Aceituno, L., Serrato, J. C., & Montilla, A. (2020). Effect of the lactose source on the ultrasound-assisted enzymatic production of galactooligosaccharides and gluconic acid. *Ultrasonics Sonochemistry, 67*, 104945.

Santos, A., Ladero, M., & Garc'ıa-Ochoa, F. (1996). Kinetic model for the hydrolysis of lactose by-galactosidase from *Kluyveromyces fragilis*. Proceedings of the seventh mediterranean congress of chemical engineering. p. 184.

Santos, A., Ladero, M., & Garcıa-Ochoa, F. (1998). Kinetic modeling of lactose hydrolysis by a *β-galactosidase* from *Kluyveromices fragilis. Enzyme and Microbial Technology, 22*(7), 558–567.

Schulz, P., & Rizvi, S. S. (2023). Hydrolysis of lactose in milk: Current status and future products. *Food Reviews International, 39*(5), 2875–2894.

Şener, N., Apar, D. K., & Özbek, B. (2006). A modelling study on milk lactose hydrolysis and β-galactosidase stability under sonication. *Process Biochemistry, 41*(7), 1493–1500.

Shakeel-Ur-Rehman. (2009). Reduced lactose and lactose-free dairy products. In P. L. H. McSweeney & P. F. Fox (Eds.), *Advanced dairy chemistry* (*Lactose, Water, Salts and Minor Constituents*) (Vol. 3, 3rd ed., p. 98e104). Springer Science and Business Media.

Shi, X., Wu, D., Xu, Y., & Yu, X. (2022). Engineering the optimum pH of β-galactosidase from aspergillus oryzae for efficient hydrolysis of lactose. *Journal of Dairy Science, 105*(6), 4772–4782.

Singh, P., Rao, P. S., Sharma, V., & Arora, S. (2021). Physico-chemical aspects of lactose hydrolysed milk system along with detection and mitigation of maillard reaction products. *Trends in Food Science & Technology, 107*, 57–67.

Singh, P., Arora, S., Rao, P. S., Kathuria, D., Sharma, V., & Singh, A. K. (2022). Effect of process parameters on the β-galactosidase hydrolysis of lactose and galactooligosaccharide formation in concentrated skim milk. *Food Chemistry, 393*, 133355.

Torres, D. P., Gonçalves, M. D. P. F., Teixeira, J. A., & Rodrigues, L. R. (2010). Galacto-oligosaccharides: production, properties, applications, and significance as prebiotics. *Comprehensive Reviews in Food Science and Food Safety, 9*(5), 438–454.

Voget, C., Borrajo, A., & Pedrazzi, C. (2022). Lactose hydrolysis in milk using a commercial recombinant β-galactosidase (lactase) from Bifidobacterium bifidum. *Food Science and Technology, 42*, e27622.

Wadiak, D. T., & Carbonell, R. G. (1975). Kinetic behavior of microencapsulated β-galactosidase. *Biotechnology and Bioengineering, 17*(8), 1157–1181.

Wang, W. W., Meng, T. T., Guo, D. Z., Hai-Le, M. A., Cao, Y., & Wang, W. X. (2015). Research progress on ultrasonic biological effect of food processing. *Science and Technology of Food Industry*.

Yang, S. T., & Okos, M. R. (1989). A new graphical method for determining parameters in Michaelis–Menten-type kinetics for enzymatic lactose hydrolysis. *Biotechnology and Bioengineering, 34*(6), 763–773.

Yin, H., Bultema, J. B., Dijkhuizen, L., & van Leeuwen, S. S. (2017). Reaction kinetics and galactooligosaccharide product profiles of the β-galactosidases from Bacillus circulans, Kluyveromyces lactis and Aspergillus oryzae. *Food Chemistry, 225*, 230–238.

Zhong, A., Katrolia, P., Zhu, C., & Kopparapu, N. K. (2023). Preparation of low lactose prebiotic milk using β-galactosidase from lotus seeds (Nelumbo nucifera). *International Journal of Food Science & Technology, 58*(5), 2609–2617.

Chapter 6
Production of Lactose Hydrolyzed Dairy Products

Shreya Panwar, Kanchanpally Saipriya, and Sonia

6.1 Introduction

Lactose, a primary sugar present in the milk of many mammals, is a type of disaccharide. It constitutes a significant portion of human milk, which is the primary source of nourishment for infants during their initial months. The lactose content (g/100 mL) of bovine milk ranges from 4.4 to 5.2 with an average concentration of around 4.8 g/100 mL anhydrous lactose and makes up approximately 6.7% of the weight of human milk which serves as the primary provider of calories in the milk (Leszkowicz et al., 2022). This molecule consists of D-glucose and D-galactose joined together by β 1, 4-glycosidic bond (Di Costanzo & Canani, 2018). Lactose molecule has a unique ability to manifest in different states and also exhibits temperature-dependent physico-chemical interrelationships (Gänzle et al., 2008). Lactose in aqueous solutions is present in α and β forms.

However, due to a lack of the enzyme lactase, some people cannot digest lactose adequately and may develop lactose intolerance, which at present affects a sizable section of the world's population. Lactose intolerance is a major problem at present across the globe, the proportion of people with the inability to digest lactose varies from 98–100% of adults in Southeast Asia to just 1% in the Netherlands (Silanikove et al., 2015). As a result, regular dairy products consumption may result in unpleasant symptoms such as bloating, gas, and diarrhoea. Lactose-hydrolysed dairy products have come to light as a potential solution for those who are lactose intolerant.

S. Panwar (✉)
Division of Crop Production, ICAR-Indian Institute of Vegetable Research, Varanasi, India

K. Saipriya
Dairy Technology Division, ICAR-National Dairy Research Institute, Karnal, India

Sonia
Department of Foods and Nutrition, CCSHAU, Hisar, India

P. S. Rao et al. (eds.), *Lactose Hydrolysis in Dairy Products*,
https://doi.org/10.1007/978-3-031-78207-7_6

The lactose free dairy has witnessed a whooping rise and is the fastest growing segment of the dairy industry. Dekker et al. (2019) also pointed out that the rate at which the lactose-free dairy industry is rising is 7.3% in comparison to the dairy industry which is 2.3%.

To understand the importance of lactose-hydrolyzed dairy products in the food business and their possible effects on human health, this chapter examines their features, production, and physico-chemical properties therefore, will give an idea about the advantages of lactose-hydrolysed dairy products on human health.

6.2 Lactose Hydrolyzed Milk

In the vast avenue of different techniques for the production of lactose free milk, the most preferred approach is lactose hydrolysis using enzyme owing to its simplicity, precision, ease of control, and economy. After attaining the desired degree of hydrolysis of the enzyme, the enzyme can be inactivated by heating. In recent advances in enzyme application, the immobilization method allows the utilisation of enzyme for longer periods. It involves the chemical or physical binding of the enzyme on a solid matrix but that renders the process expensive. The physical immobilization method involves the usage of an ultrafiltration membrane (hollow fibres) that separates substrate from enzyme solution. However, Novalin et al. (2005) reported the microbial growth in enzyme solution that was decreased by using sterile filtration and UV irradiation.

6.2.1 Production

The lactose hydrolysis in milk is affected by a combination of factors such as pH, temperature and enzyme concentration. Lactose hydrolysis occurs maximum at 6.5 pH and till pH 5.5 hydrolysis will occur but with decreasing the pH below 5.5 the enzyme will likely get inactivated. Popescu et al. (2021) used temperatures from 4 to 70 °C and maximum hydrolysis was observed at temperatures 4–6 °C for 30 h and 38–40 °C for 5 h. Enzyme concentrations of 0.1, 0.15 and 0.3% were used and maximum hydrolysis was observed with 0.3% enzyme concentration. Choi et al. (2007) developed the lactose hydrolyzed (LH) milk. In this study, raw milk was treated with 0.03% enzyme for 24 h and 0.1% enzyme for 40 h to hydrolyze lactose partially and completely at 4 °C. Hydrolyzed milk was heated to 72 °C for 5 min to inactivate the enzyme and cooled to 45–50 °C and then subjected through nanofiltration to retain glucose and galactose, thereby reducing the sweetness of the milk. The monosaccharides obtained during lactose hydrolysis are sweeter than lactose and during processing it leads to off flavours via maillard reaction. Messia et al. (2007) prepared pasteurized micro-filtered lactose hydrolyzed skim milk by treating skim milk with enzyme 0.4 mL/L at 10 °C for 21 h. The treated milk was micro-filtered and pasteurized and it was observed that the treated milk had an extended shelf life of 21–27 days at refrigeration storage. Whey protein nitrogen index (WPNI) can be used as a thermal marker for indicating heat damage in lactose hydrolysed milk. Neves and de Oliveira (2020) studied the hydroxyl methyl furfural (HMF), lactulose and WPNI of commercial lactose hydrolysed (LH) UHT milk and observed that the HMF content of UHT milk was significantly higher in LH UHT milk 42.15 μmol/L than normal UHT milk 13.11 μmol/L, whereas lactulose content was significantly lower in LH UHT milk (13.03 mg/100 mL compared to normal UHT milk 35.59 mg/100 mL). However, no significant difference was observed between WPNI of LH UHT and normal milk.

6.2.2 Properties

The resultant products of lactose hydrolysis during LH milk production are more reactive in the maillard reaction and produce stable products. These resultant maillard reaction products have both positive and negative effects. Though melanoidins possess antioxidant and antimicrobial activities owing to their chelation property, hydroxyl methyl furfural (HMF) and advanced glycation end products have been reported to possess toxic and carcinogenic properties. Therefore, the reduction of these products in LH milk is important. It has been reported that production of these components was reduced by inhibiting the initial reactants by the addition of enzymes such as fructosamine oxidase, fructosamine kinase, glucose oxidase and hexose oxidase. The mechanism involves the incorporation of functional ingredients such as flavonoids and polyphenols and the modification of monosaccharides by transgalactosylation (Singh et al., 2021). Jensen et al. (2015) reported that during 4-month storage study of LH UHT milk, increase in bitterness was observed, which was attributed to the presence of primary amines, which are present more in LH milk.

6.3 Lactose Hydrolyzed Yoghurt

Yoghurt is one of the oldest fermented products and possesses smooth viscous gel with a sharp acid taste and green apple flavour. It is generally produced by *Streptococcus thermophilus* and *Lactobacillus bulgaricus* (1:1). During fermentation, the lactose is utilized by the bacteria to produce lactic acid which results in bringing the pH below 5.3 thereby, resulting in irreversible protein destabilization and flavour formation (Das et al., 2019). During the fermentation, the lactose gets converted to lactic acid, milk fat is broken down to fatty acids and proteins get hydrolysed to free amino acids and peptides. Further, various metabolites of lactose including lactic acid, acetic acid, diacetyl, folic acid and formic acid are generated by the action of bacteria during fermentation which also results in reduced pH of the environment. Further, the reduced pH promotes the growth of *Lactobacillus delbrueckii* subsp. *bulgaricus*, where it hydrolyses the casein into free amino acids and peptides and stimulates the growth of the *Streptococcus thermophilus* (Jia et al., 2021). Therefore, symbiotic association and proto-cooperation between *Lactobacillus delbrueckii* subsp. *bulgaricus* (LB) and *Streptococcus thermophilus* (ST) provide the conducible environment for the fermentation process to complete. LB and ST possess lactase activity (Sarkar, 2006). For low lactose or lactose-free yoghurt, the milk is treated with β-galactosidase enzyme before (pre-hydrolysis) or during (co-hydrolysis) fermentation process. Several factors affect the concentration of lactose, the physicochemical and sensory properties of yoghurt such as the level of enzyme, time of addition, and storage conditions (Vénica et al., 2013).

6.3.1 Production

There are two methods to produce lactose hydrolysed yoghurt: Pre-hydrolysis and Co- hydrolysis. In pre-hydrolysis, the enzyme is added and incubated as per the manufacturer's details. Then milk is subjected to pasteurization and after that, it is cooled to fermentation temperature. In the case of co-hydrolysis, the enzyme is added to the starter culture and it gets inactivated completely when the product reaches a pH of 5.5 usually after 2–3 h (Zagorska et al., 2020). Higher lactose hydrolysis has been observed in pre-hydrolysis (>90%) than in co-hydrolysis (>80%). However, from an economic point of view, co-hydrolysis is considered an optimal method as it reduces the production time compared to pre-hydrolysis. Popescu et al. (2021) reported the reduction in fermentation time in pre-hydrolysis and co-hydrolysis by 30 and 60 min respectively. The rationale behind accelerated fermentation time is the presence of the monosaccharides as they are easily fermented causing the higher growth of lactic acid bacteria.

Wolf et al., (2018) and Vénica et al. (2013) reported that for the production of the Lactose hydrolyzed (LH) yoghurt, β-galactosidase enzyme was added as per manufacture recommendations and incubated for 1 h, followed by the addition of the starter culture. Incubation was done till pH reached 4.5 ± 0.20. It was found that the optimum pH range for β-galactosidase enzyme was 5.4 to 8.0 as below pH 5.4 pH enzyme would likely get inactivated. It was also reported that lactose content decreased by 75 to 78% when the enzyme was added to the starter culture. Nagaraj et al. (2009) pasteurized and cooled the milk to 40 °C followed by the addition of enzymes at different concentrations from 0.12 to 0.56%. It was then incubated for pre-hydrolysis. After that, the milk was heated to 75 °C for 2 min to inactivate the enzyme. The enzyme concentrations of 0.16, 0.32 and 0.52% were reported to have 50, 70 and 90% hydrolysis. Also, 50 and 70% LHY found higher scores for body and texture and overall acceptability than control. On the other hand, 90% of LHY had lower sensory scores. It was also observed that with hydrolysis the setting time was reduced by 35–40 min. The penetration and syneresis reportedly increased as hydrolysis increased in the pre-hydrolysis and co-hydrolysis methods (Sharanagouda et al., 2022).

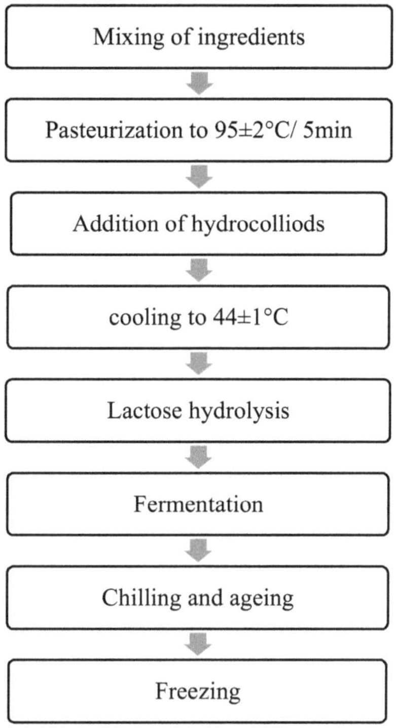

Fig. 6.1 Schematic flowchart chart for preparation of frozen yoghurt. (Skryplonek et al., 2019)

Another very popular variant of yoghurt is frozen yoghurt/Yoghurt ice cream, the preparation of which has been presented in Fig. 6.1, for the production of frozen yoghurt Skryplonek et al. (2019) calculated amounts of whole milk, cream and skim milk powder to get a mix of 5 and 12% MSNF. Sucrose was added @6.5% of the mix and the mix was heated to 95 ± 2 °C/5 min. Hydrocolloids were also added to improve the melting resistance and texture of the final product.

The mix was cooled to 44 ± 1 °C and the enzyme was added and incubated for 30 min. The added enzyme was expected to get inactivated at pH 5.5. After that, the starter culture was added and incubated until the pH reached 4.6 ± 1 °C. Then yoghurt was cooled to room temperature for about 3 h and aged overnight under refrigeration conditions. The mix was then frozen, packed and stored at −18 ± 1 °C. They observed that the lactose content was initially 4.4% and it was reduced to 1.5% after 20 min incubation with enzyme and after 120 min it was undetectable. However, due to the lower fat content, the viscosity, creaminess, hardness, flavour and melting rate are affected. To avoid this, stabilizers are added as they improve viscosity, incorporation of air and retards melting. Carrageenan and corn starch were reported to be used in lactose-free yoghurt. Carrageenan forms complexes with casein and improves the structure and stability of the product.

For the production of drinkable yoghurt Vénica et al. (2013) standardised milk to 3% fat and 8% sugar in sweetened yoghurt. The milk was heated to 90 ± 2 °C and cooled to 42 ± 2 °C. Then enzyme and starter culture were added and incubated till the pH reached 4.7 ± 0.10. During 28 days of storage, there was no significant change in the lactose hydrolysis due to the inactivation of enzymes during fermentation. A significant change was not observed in the pH and acidity of yoghurt during 28 days of storage. It was also observed that the syneresis of sweetened and natural yoghurt significantly increased from 0 to 8.2% and 0 to 10.8% respectively during 28 days of storage. This was attributed to the presence of a higher level of soluble sugars in comparison with the untreated samples, as hydrolysed samples are sweeter.

6.3.2 Properties

Yoghurt contains essential nutrients, highly bioavailable proteins, calcium, potassium, phosphorous, magnesium, riboflavin, vitamins B_{12} and vitamin A. It contains high levels of water-soluble vitamins such as B1, B2, B9 and B12, minerals such as calcium, magnesium, phosphorous and zinc. Lactose hydrolysis yields glucose and galactose that give elevated sweetness. Lactose hydrolysed yoghurt tasted sweeter with a light caramel flavour, and viscosity and syneresis were higher compared to non-hydrolysed yoghurt (Popescu et al., 2021). The control sample (no enzyme) was less sweet than enzyme-treated samples (Zagorska et al., 2020). Also, during hydrolysis of lactose, galactooligosaccharides will be produced, which are prebiotics and may help in better survivability of lactic acid bacteria and may also act as a fat replacer and can help in improving the texture.

6.4 Lactose Hydrolyzed Ice-Cream

Ice cream structure consists of three phases fat globules, air bubbles absorbed interfacial material and ice crystals surrounded by a freeze-concentrated aqueous serum or matrix phase that contains the sugars, proteins, polysaccharides and salts. It has been reported that at ambient temperature, the solubility of lactose is ten times lower than that of sucrose, thereby lactose crystallization occurs. Owing to the lactose hydrolysis, monosaccharides such as glucose and galactose are produced. These monosaccharides cause depression in the freezing point and improve the texture of ice cream (Özdemir et al., 2018).

6.4.1 Production

Matak et al. (2003) formulated LH vanilla ice cream consisting of 10% milk fat, 11% MSNF, 15% sucrose and 0.3% stabilizer. The enzyme was added to the ice cream mix and stored for 24 h at 4 °C. Before freezing the hydrolysed mix was mixed with the non-hydrolysed mix to produce different levels of hydrolysis such as

Table 6.1 Lactose hydrolyzed ice cream preparation

Product	Enzyme	Hydrolysis	References
Lactose-reduced ice cream	Milk treated with 0.5 g/L and incubated at 37 °C for 4 h and inactivated by heat treatment	86–97%	Tsuchiya et al. (2017)
Effect of hydrolysis of lactose and sucrose on firmness of ice cream	Ice cream mix was pasteurized and cooled to 4.5 °C Lactase-7.5 mL/100 L mix and incubated for 48 h	78.4%	Lindamood et al. (1989)
Influences of lactose hydrolysis of milk and sugar reduction on some physical properties of ice cream	Pasteurized skim milk Enzyme-1.5 mL/L Incubation-15,35,75 and 240 min	Lactose hydrolysis-25, 50. 75 and more than 75%	Abbasi and Saeedabadian (2015)

0, 80 and 100% and frozen with 75% overrun. The freezing point of the ice cream mix of control and 80 and 100% lactose hydrolysis were −1.61–2.18 and −2.22 °C respectively. The yield stress of control ice cream was significantly different with 80 and 100% hydrolysis and there was no significant difference with 80 and 100% hydrolysis. The yield stress was reported to decrease with the increase in the hydrolysis, which accounted for the increased softness of the lactose-hydrolysed ice cream produced. Guy (1980) utilized the sweet whey solids in ice cream, where the enzyme was added to raw milk at 3 °C for 16 to 18 h and the milk was pasteurized and used for cheddar cheese production. The hydrolysed sweet whey (67 or 79% hydrolysis) was concentrated at 49% TS. The lactose hydrolysed whey substitution of 2.75% to that of non-fat milk solids was produced of comparable quality to the control (Guy, 1980). A few pieces of literature regarding lactose-hydrolyzed ice cream have been presented in Table 6.1.

6.4.2 Properties

The physico-chemical properties of ice cream prepared with lactose hydrolysis have no significant difference concerning the control. However, hydrolysis leads to a decrease in the freezing point of the ice cream mix. The freezing point of the ice cream mix of control and 80% and 100% lactose hydrolysis as reported by Matak et al. (2003) were −1.61 °C and −2.18 °C and −2.22 °C respectively. Lindamood et al. (1989) also reported a decrease in the freezing point of ice cream from −1.42 to −1.92 °C and it was also observed that the relative sweetness increased from 16 to 18.5, firmness decreased and the rate of melting increased from 0.45 ± 0.13 to 0.81 ± 0.07 mL/min. Özdemir et al. (2018) reported lower overrun, viscosity and higher melting rate than control in ice cream prepared using Lactose-free UHT milk. Abbasi and Saeedabadian (2015), reported depression in the freezing point (−0.3 °C), it was concluded that the increased freezing point depression was due to the soluble constituents of low molecular weight salts and sugars. The viscosity was significantly increased with the increase in the lactose and till 25% LH there was no significant difference when compared to the control. Hardness was significantly increased with LH as it was 7.1 N in control and 46 N in >75% LH. Melting resistance was decreased with an increase in the LH (1.04 in control and > 75% LH 1.34 mL/min).

6.5 Lactose Hydrolyzed Milk Powder

Drying is a preservation method to increase the shelf life by removing moisture from food by vaporization (spray-drying) or sublimation process (freeze-drying), which reduces the microbial growth and chemical, enzymatic reactions in the food (Guine, 2018). Normally roller drying, spray drying, and freeze-drying methods are used for drying milk-based products and the selection of dryer mostly depends on the type of raw material used and its final application. Roller-dried powder has good porosity and rehydration properties but high temperature results in protein denaturation and cooked flavour. Compared to spray drying, roller drying needs less space for operation and thus, it is a more economical process (Anandharamakrishnan, 2017). Spray drying is a rapid moisture removal process that converts the liquid into powder by the process of atomization in the presence of a hot drying medium and leads to the amorphous lactose formation as a continuous phase and remaining components as dispersed phase such as fat globules, proteins and air cells (Shrestha et al., 2007). Spray drying has been reported to have an encapsulation effect, thus helping to advance the functionality of dairy products. Freeze drying is a non-thermal process using low temperatures for the removal of water from a product by sublimation and desorption process. The product obtained using freeze drying has a long shelf life, low moisture content, rapid rehydration, and no nutrition damage but this process is slower and involves a higher cost of production. It is usually done in batch process thus making it an expensive affair. The alternate process for freeze-drying is spray drying in terms of cost-effectiveness (Guine, 2018). The presence of a significant number of people having lactose intolerance reduces the consumption of milk and milk products. The production of lactose-reduced products helps enhance the consumption of dairy products (Shrestha et al., 2007). As per the literature, a spray drier and freeze drier were used for producing lactose hydrolysed milk powder. Hydrolysis of dried powder affects the quality like decreased shelf life and glass transition temperature, increased darkening and speed of maillard reaction (de Paula et al., 2021).

6.5.1 *Production*

For the preparation of lactose hydrolysed dried milk powder (LHDMP), lactase enzyme was added to 40% total solids concentrated milk and incubated at 7 ± 2 °C for 24 h for lactose hydrolysis. Then it was spray dried using inlet air temperature

(150–160 °C), outlet air temperature (85–90 °C) and airflow (40–50 L/min). The final dried product was stored in different storage and package conditions such as plastic packaging and styrofoam boxes without air inlet and light for 21 days at 21 °C, desiccators with different relative humidity (11.1, 33.1, 43.2, 54.4, and 75.5%) for 9 weeks and controlled temperature of 50 ± 1 °C with less than 10% relative humidity for 21 days. The obtained powder having higher humidity than the control due to the lower glass transition, will eventually affect difficulty in the drying. Due to this, the hygroscopicity nature of the powder was increased. The moisture of the control sample was 4.1 g/100 g and lactose hydrolysed powder (99% hydrolysis) was 10.1 g/100 g. The fat content of LHDMP is 30%. Compared to the LHDMP, the control sample had higher free fat content owing to its transformation of amorphous lactose to crystalline form affects the stability of the powder and also affects the flavour change, increased viscosity and non-enzymatic browning. It retains a higher amount of water per 100 g of solids. LHDMP contains a higher free HMF content of 30 µm/L than control 18 µm/L due to the lactose hydrolysis will increase the reducing sugars glucose and galactose and will enhance the non-enzymatic browning. High temperature of storage of LHDMP was also observed with high free HMF content, where it enhances the non-enzymatic browning reaction, darker colour values such as lower L* and higher a* and b* values and affects the solubility (de Paula et al., 2021). Fernández et al., 2003 observed the non-enzymatic browning after 4 weeks of storage at 37 °C due to the presence of higher reactive molecules such as glucose and galactose were increased. Rodrigues et al., 2020 was prepared lactose hydrolysed whole milk powder by treating the enzyme at 40 ± 1 °C, for 2 h. Then it was heat treated to 95 °C/60 s and dried using a spray drier. Lactose hydrolysed skim milk was manufactured from concentrated skim milk of 40% total solids. To that lactase enzyme was incubated for 4 h (Fialho et al., 2018a, b). Fialho et al., 2018a, used different inlet air temperatures from 115 to 160 °C and flow rates from 0.3 to 1.5 Kg/h for spray drying of lactose hydrolysed milk powder. Among different inlet air temperatures and flow rates of spray drier, the acceptable powder was obtained using an inlet air temperature of 145 °C and flow rate of 1 Kg/h in terms of physic-chemical and techno-functional properties (Fialho et al., 2018a, b). The accelerated storage of LH milk powder at 60 °C observed with increased aggregation and decreased rehydration capacity and protein glycation capacity causes the development of monosaccharides protein interaction leads to the maillard reaction (Fialho et al., 2021). With the increase in the hydrolysis, the significant increase in the moisture was observed. Production of lactose hydrolysed milk powder by spray drying is difficult as at lower inlet and outlet temperatures also it sticks to the wall of the drier (Torres et al., 2017). Queiroz et al. (2021) reported that LH goat milk powder influences the darkening and free fat content. Shrestha et al. (2007) manufactured lactose hydrolysed milk powder using inlet air temperature 130 and outlet air temperature 65 °C. About 50% of the solids were recovered from the spray drier, 25% of the solids were recovered from the cyclone and the rest of the solids were removed manually by sweeping. They observed that the stickiness behaviour of the powder makes it difficult to recover the powder from the drier. The LHP absorbs moisture quickly and in 24 h storage the

Table 6.2 Different processing conditions for manufacturing lactose hydrolysed milk powders

Milk type	Enzyme treatment	Processing condition	References
Low fat milk (0.5% fat)—UHT	40 mg/L and incubated for 1 h at room temperature	Inlet air temperature (IAT)—170 °C Outlet air temperature (OAT)—70 °C	San Jose et al. (1977)
Non-fat milk (0.05%)—HTST pasteurization	300 mg/L and incubated for 48 h at 4 °C	IAT—170 °C OAT—70 °C	San Jose et al. (1977)
Whole concentrated milk—40% total solids	Lactase enzyme was inactivated by heating to 90 to 95 °C for 60 s	IAT—170 ± 5 °C OAT—85 ± 5 °C	Torres et al. (2017)
Hydrolyzed goat milk	–	IAT—160 ± 1 °C OAT—80 ± 4 °C	Queiroz et al. (2021)
Lactose hydrolysed milk	–	Concentrate (30° brix)—Rising film vacuum evaporator IAT—130 °C OAT—65 °C	Shrestha et al. (2007)

LHP was not crystallized but the SMP was crystallised at water activity above 0.753 (Shrestha et al., 2007). Different authors prepared lactose hydrolysed milk powders using different processing conditions given in Table 6.2.

6.5.2 Physicochemical Properties

Properties of dried milk powders such as glass transition temperature and crystallization are important as they affect the product quality such as stickiness, caking, porosity, solubility, dispersibility and flowability (Torres et al., 2017). It was observed that storage of milk powders at a temperature higher than the glass transition temperature, will lead to the transition of amorphous particles from glassy to rubbery state. The glass transition temperature is affected by the molecular weight, the process of lactose hydrolysis produces low molecular weight monosaccharides such as glucose and galactose which decreases the glass transition temperature of powder than the normal milk powder (Fernández et al., 2003). The presence of amorphous lactose and other monosaccharides above their glass transition temperature leads to the formation of a rubbery state owing to the hygroscopic nature of the amorphous nature of sugars. Further, it leads to stickiness, caking and crystallization. Eventually affects the yield, operation, handling and storage difficulties. Milk powders containing monosaccharides like lactose hydrolysed milk powder are thereby susceptible to caking or loss of flowability due to their moisture sorption or exposure to higher temperatures. These affect the rehydration and dispersibility properties (Fernández et al., 2003). The fat oxidation and non-enzymatic browning may also decrease the quality of milk powder during storage. The glass transition temperature (Tg) of freeze-dried skim milk powder was reported −50 °C and for freeze-dried lactose hydrolyzed (LH) skim milk powder was reported −65 °C. The water activity of freeze-dried LH powder was 0.16. During storage, a slight increase in water content was observed, which increases the stickiness and caking drastically (Jouppila & Roos, 1994). Shrestha et al. (2007) reported the glass transition temperature of lactose hydrolyzed whole milk powder at 58 °C, lactose hydrolysed skim milk powder at 48 °C and normal skim milk powder at 93 °C. Lactose hydrolysed milk powder water activity below 0.2 aw is lower than normal powder owing to its decrease in the amount of nonsolvent water for a protein (San Jose et al., 1977).

6.6 Lactose Hydrolyzed Prebiotic Whey

The by-product of the cheese making process, termed whey, has conventionally been measured as an unwanted product with less commercial significance (Ayed et al., 2023). The composition and characteristic features of the whey mainly depend upon the quality of the milk used for processing, production technology and the method of processing used for the end product. The liquid form of whey mainly consists of approximately 93 per cent of water. It contains about half of the total solids present in the natural milk used for the processing of the cheese. In particular, lactose is the main component among the total solids in the whey beverage with 1% of the proteins (Mudgil & Barak, 2019). However, the non-digestible carbohydrates, prebiotics, are the fermented ingredients which help to induce the growth and activity of the beneficial micro-flora in the gastrointestinal tract and are also commonly used to promote the growth of the probiotic bacteria, providing advantages for the host's health and well-being (Davani-Davari et al., 2019). Prebiotic beverages prepared from whey have also recently gained popularity as a healthy meal alternative (Ayed et al., 2023).

6.6.1 Production

Various types of prebiotics, including inulin, galacto-oligosaccharides, fructo-oligosaccharides and xylo-oligosaccharides are found in whey-based prebiotic beverages. These beverages are prepared with whey and possess a variety of physiochemical qualities, depending upon the prebiotic type and formulation. For instance, oat beta-glucan and whey protein isolates improved the taste, usability, and stability of an orange juice beverage (Wan & Xu, 2018). In the addition of xylo-ligosaccharide (XOS) to strawberry-flavoured whey beverages, the outcome was comparable (Souza et al., 2019). The use of lactose-hydrolyzed whey in beverage production is another encouraging trend. An acceptable-grade carbonated lemon whey beverage was produced using lactose hydrolyzed whey, lemon juice (4.5%), sugar (10%), ginger juice (1%) and salt (0.6%) and was shelf stable for up to 49 days at a temperature of 7 °C (Patel, 2017). Another study by Atallah and Gemiel (2020) was done for the manufacturing of carbonated whey drinks using lactose-hydrolyzed whey, sugar, fruit juices, and/or herbs. Dutra Rosolen et al. (2015) found that during lactose hydrolysis, at the temperature of 10 °C, the enzyme *K. lactis β-galactosidase* was more effective in the cheese whey and milk as compared to *A. oryzae.*

6.6.2 *Properties*

Owing to their high nutritional value and good amounts of proteins, the lactose hydrolysed whey beverages are a great source of energy and nutrients. Branched-chain amino acids (BCAA) such isoleucine, leucine and valine are found in ample amounts in the whey proteins. As compared to other amino acids, BCAAs are utilised first during the time of physical activity and resistance training because they are metabolised directly into the muscle tissues (Sherwood & Jenkins, 2007). The whey protein fractions contain glycomacropeptide (GMP) which is produced during cheese making using rennet, alpha-lactalbumin a calium-binding protein and lactoferrin, an iron-binding protein. Owing to the presence of lactoferrin, the lactose hydrolysed whey beverages can be used as functional foods to increase the absorption of iron from food and also helps to prevent various infections from adhering to the intestinal walls. It is crucial for the nutrition of infants and children. Moreover, these beverages can enhance calcium absorption, which is essential for the elderly population suffering from osteoporosis (Jeličić et al., 2008).

6.7 Lactose Hydrolysed Cheese

Potable milk is the largest category of lactose-free dairy products, while lactose free cheese is the fastest growing product in this segment of the dairy industry, and it is expected to grow at a pace of 8.4% (Dekker et al., 2019). Cheese by its innate nature is a low-lactose-containing dairy product, as its processing involves the removal of whey and most of the lactose is excreted along with whey (Panseri et al., 2021). Also, some cheese like Gouda cheese processing involves a specific step of curd washing which also brings down the lactose level (Walstra et al., 1999). In the case of ripened hard cheese like Parmesan, Cheddar or Swiss cheeses and semi-hard cheese the lactose continues to be converted into lactic acid by the action of lactic acid bacteria, rendering them naturally lactose-free (Panseri et al., 2021). In the case of cheese, only the fresh cheese/young cheese may have a chance of encompassing lactose levels that can be difficult to be tolerated by the lactose intolerant population.

6.7.1 Production

Cheese can be rendered lactose-free by introducing lactase to the cheese milk before initiating the renneting process. This procedure is particularly advantageous for fresh, youthful cheeses that are recognized for their notable lactose content, and at present, lactase enzymes are employed for this specific purpose. In the case of matured cheeses, the lactic acid bacteria present will have already metabolized all the lactose, rendering lactase incubation unnecessary. Historical literature often references the treatment of cheese milk with lactase as a means to enhance acidification during the cheese production process (Marschke et al., 1980). Moreover, the addition of lactase is acknowledged for its potential to amplify the development of cheese flavours during the ageing process. Nonetheless, the exact attribution of these effects remains somewhat unclear—whether they arise from the encouragement of microbial growth within the cheese due to lactose hydrolysis as microbes readily grow on glucose sugar rather than galactose or lactose, or from any residual proteolytic activity inherent in the commercially available lactase preparations. It is plausible that the liberation of amino acids from milk proteins could stimulate the activity of the cheese culture in these experimental scenarios (Dekker et al., 2019).

6.7.2 Properties

Marschke and Dulley (1978) reported that the cheese, prepared using milk that had undergone 70% lactose hydrolysis exhibited texture, flavour, and overall body similar to cheese aged for 6–8 months despite being aged for only 3–4 months. It was also reported that the whey derived from cheese produced using hydrolysed milk showed potential as a sweetener. Raza et al. (2021) also reported a similar appearance and overall acceptability of low-lactose (0.8 g/100 g) cheese containing prebiotics produced using exogenous lactose which converted lactose into galacto-oligosachharaide.

6.8 Lactose Hydrolysed Infant Formula

Infant formula often encompasses lactose as a principal component produced from cheese or casein whey, also the synthetic derivative of lactose, lactulose, is considered as a second-line drug in the treatment of constipation owing to its osmotic properties. It has been also reported that lactulose plays an important role in infant nutrition as a "Bifidus Factor" that assists in the growth of *Bifidobacterium* in the intestinal tract and thereby helps in improving the gut microbiota of the infants (Leszkowicz et al., 2022). However, few infants are innately susceptible to lactose intolerance due to congenital lactase deficiency and galactosemia. Therefore, lactose-free formulas have emerged as an alternative to soy formula for parents wishing to avoid lactose in the infant formula O'Connor (2009).

6.8.1 *Production*

Lactose free infant formula are produced using lactose hydrolysed milk. Torres et al. (2017) reported the production of lactose-free infant formula by using Concentrated whole milk that was obtained from the reconstitution of milk powder in water at 25 °C and contained approximately 40% total solids and was subjected to hydrolysis using 0.2% w/w of Lactomax Super enzyme by Prozyn®.

6.8.2 *Properties*

Torres et al. (2017) concluded production and storage of lactose hydrolyzed milk powder or low lactose powdered milk (LLPM) using the spray drying technology is very difficult as most of the powder tends to stick to the insides of the dryer even at very low inlet/outlet air temperatures. Issues of caking and darkening have also been reported in the production of LLPM. These issues often lead to low production yield, operational problems and difficulty in powder handling. Schuck et al. (2015), indicated that an increased number of molecules in terms of glucose and galactose in its amorphous state as compared to lactose upon hydrolysis contributes to high hygroscopicity, which renders processing of LLPM difficult due to increased stickiness that causes clogging.

6.9 Conclusion

Lactose intolerance is a widespread condition affecting a significant portion of the global population. However, Dairy enterprises and manufacturers have improved processes and technologies, using immobilized enzymes, genetic engineering, membrane filtration, and fermentation to remove lactose to create dairy products that are more accessible to individuals with lactose intolerance. This enzymatic process is crucial for producing dairy products that are not only more easily digestible but also maintain the taste and texture of traditional dairy items. Various methods and technologies employed in the hydrolysis of lactose, have evolved to enhance the quality of lactose-free dairy products. The other important factor of lactose-hydrolysed products is their ability to provide essential nutrients while catering to lactose-intolerant individuals. The market for these products continues to grow, with an increasing demand for lactose-free alternatives that taste as good as conventional dairy products. In the ever-evolving landscape of the food industry, lactose-hydrolyzed dairy products offer a bridge between the needs of lactose-intolerant consumers and their desire for delicious, nutritious dairy items. As technology and consumer preferences continue to evolve, this chapter has laid a strong foundation for understanding the production and potential benefits of lactose-hydrolyzed dairy products in meeting the diverse dietary requirements of the modern world.

References

Abbasi, S., & Saeedabadian, A. (2015). *Journal of Food Science and Technology, 52*, 367–374.

Anandharamakrishnan, C. (Ed.). (2017). *Handbook of drying for dairy products*. Wiley.

Atallah, A. A., & Gemiel, D. G. (2020). Preparation of new carbonated beverages based on hydrolyzed whey by fruit and some herbs. *American Journal of Food Science and Technology, 8*, 49–55.

Ayed, L., M'hir, S., & Asses, N. (2023). Sustainable whey processing techniques: Innovations in derivative and beverage production. *Food Bioscience, 53*, 102642.

Choi, S. H., Lee, S. B., & Won, H. R. (2007). Development of lactose-hydrolyzed milk with low sweetness using nanofiltration. *Asian-Australasian Journal of Animal Sciences, 20*(6), 989–993.

Das, K., Choudhary, R., & Thompson-Witrick, K. A. (2019). Effects of new technology on the current manufacturing process of yogurt-to increase the overall marketability of yogurt. *Lwt, 108*, 69–80.

Davani-Davari, D., Negahdaripour, M., Karimzadeh, I., Seifan, M., Mohkam, M., Masoumi, S. J., Berenjian, A., & Ghasemi, Y. (2019). Prebiotics: Definition, types, sources, mechanisms, and clinical applications. *Food, 8*(3), 92.

de Paula, I. L., Teixeira, E. B. S., Francisquini, J. D. A., Stephani, R., Perrone, Í. T., de Carvalho, A. F., & de Oliveira, L. F. C. (2021). Buffalo powder dairy products with and without lactose hydrolysis: Physical-chemical and technical-functional characterizations. *LWT, 151*, 112124.

Dekker, P. J., Koenders, D., & Bruins, M. J. (2019). Lactose-free dairy products: Market developments, production, nutrition and health benefits. *Nutrients, 11*(3), 551.

Di Costanzo, M., & Canani, R. B. (2018). Lactose intolerance: Common misunderstandings. *Annals of Nutrition and Metabolism, 73*(4), 30–37.

Dutra Rosolen, M., Gennari, A., Volpato, G., & Volken de Souza, C. F. (2015). Lactose hydrolysis in milk and dairy whey using microbial β-galactosidases. *Enzyme Research, 2015*, 806240.

Fernández, E., Schebor, C., & Chirife, J. (2003). Glass transition temperature of regular and lactose hydrolyzed milk powders. *LWT-Food Science and Technology, 36*(5), 547–551.

Fialho, T. L., Martins, E., Silva, C. R. D. J., Stephani, R., Tavares, G. M., Silveira, A. C. P., & de Carvalho, A. F. (2018a). Lactose-hydrolyzed milk powder: Physicochemical and technofunctional characterization. *Drying Technology, 36*(14), 1688–1695.

Fialho, T. L., Martins, E., Silveira, A. C. P., Silva, C. R. D. J., Perrone, Í. T., Schuck, P., & de Carvalho, A. F. (2018b). Lactose hydrolyzed milk powder: Thermodynamic characterization of the drying process. *Drying Technology, 36*(8), 922–931.

Fialho, T. L., Nascimento, L. G. L., Moreau, A., Delaplace, G., Martins, E., Perrone, Í. T., & Junior, P. P. D. S. P. (2021). Sugar type matters in spray drying II: Glycation effects on physicochemical characteristics of aged lactose-hydrolyzed milk powder. *Food Structure, 30*, 100215.

Gänzle, M. G., Haase, G., & Jelen, P. (2008). Lactose: Crystallization, hydrolysis and value-added derivatives. *International Dairy Journal, 18*(7), 685–694.

Guine, R. (2018). The drying of foods and its effect on the physical-chemical, sensorial and nutritional properties. *International Journal of Food Engineering, 2*(4), 93–100.

Guy, E. J. (1980). Partial replacement of non fat milk solids and cane sugar in ice cream with lactose hydrolyzed sweet whey solids. *Journal of Food Science, 45*(1), 129–133.

Jeličić, I., Božanić, R., & Tratnik, L. (2008). Whey based beverages-new generation of dairy products. *Mljekarstvo, 58*(3), 257–274.
Jensen, S., Jansson, T., Eggers, N., Clausen, M. R., Larsen, L. B., Jensen, H. B., & Bertram, H. C. (2015). Storage-induced changes in the sensory characteristics and volatiles of conventional and lactose-hydrolyzed UHT processed milk. *European Food Research and Technology, 240*, 1247–1257.
Jia, W., Liu, Y., & Shi, L. (2021). Integrated metabolomics and lipidomics profiling reveals beneficial changes in sensory quality of brown fermented goat milk. *Food Chemistry, 364*, 130378.
Jouppila, K., & Roos, Y. H. (1994). Glass transitions and crystallization in milk powders. *Journal of Dairy Science, 77*(10), 2907–2915.
Leszkowicz, J., Plata-Nazar, K., & Szlagatys-Sidorkiewicz, A. (2022). Can lactose intolerance be a cause of constipation? *A Narrative Review. Nutrients, 14*(9), 1785.
Lindamood, J. B., Grooms, D. J., & Hansen, P. M. T. (1989). Effect of hydrolysis of lactose and sucrose on firmness of ice cream. *Food Hydrocolloids, 3*(5), 379–388.
Marschke, R. J., & Dulley, J. R. (1978). The effect of partial lactose hydrolysis on the manufacture and ripening of Cheddar cheese. *Australian Journal of Dairy Technology, 33*(4), 139.
Marschke, R. J., Nickerson, D. E. J., Jarrett, W. D., & Dulley, J. R. (1980). A cause of increased proteolysis in cheddar cheese manufactured from milk containing added Maxilact. *Australian Journal of Dairy Technology, 35*, 84–88.
Matak, K. E., Wilson, J. H., Duncan, S. E., Wilson, E. J., Hackney, C. R., & Sumner, S. S. (2003). The influence of lactose hydrolysis on the strength and sensory characteristics of vanilla ice cream. *Transactions of the ASAE, 46*(6), 1589.
Messia, M. C., Candigliota, T., & Marconi, E. (2007). Assessment of quality and technological characterization of lactose-hydrolyzed milk. *Food Chemistry, 104*(3), 910–917.
Mudgil, D., & Barak, S. (2019). Dairy-based functional beverages. In *Milk-based beverages* (pp. 67–93). Woodhead Publishing.
Nagaraj, M., Sharanagouda, B., Manjunath, H., & Manafi, M. (2009). Standardization of different levels of lactose hydrolysis in the preparation of lactose hydrolyzed yoghurt. *Iranian Journal of Veterinary Research, 10*(2), 132–136.
Neves, L. N. D. O., & de Oliveira, M. A. L. (2020). Effects of enzymatic lactose hydrolysis on thermal markers in lactose-free UHT milk. *Journal of Food Science and Technology, 57*, 3518–3524.
Novalin, S., Neuhaus, W., & Kulbe, K. D. (2005). A new innovative process to produce lactose-reduced skim milk. *Journal of Biotechnology, 119*(2), 212–218.
O'Connor, N. R. (2009). Infant formula. *American Family Physician, 79*(7), 565–570.
Özdemir, C., Arslaner, A., Özdemir, S., & Uğurlu, G. (2018). Ice-cream production from Lacose-free UHT Milk. *Journal of Food Science and Engineering, 8*, 210–214.
Panseri, S., Pavlovic, R., Castrica, M., Nobile, M., Di Cesare, F., & Chiesa, L. M. (2021). Determination of carbohydrates in lactose-free dairy products to support food labelling. *Food, 10*, 1219.
Patel, R. (2017). Technology for carbonated lemon whey beverage. *Research & Reviews. Journal of Food and Dairy Technology, 5*(2), 30–37.
Popescu, L., Bulgaru, V., & Siminiuc, R. (2021). Effect of temperature, pH and amount of enzyme used in the lactose hydrolysis of milk. *Food and Nutrition Sciences, 12*(12), 1243–1254.
Queiroz, E. S., Rezende, A. L. L., Perrone, Í. T., Francisquini, J. D. A., de Carvalho, A. F., Alves, N. M. G., et al. (2021). Spray drying and characterization of lactose-free goat milk. *LWT, 147*, 111516.
Raza, A., Iqbal, S., Shah, F.-U.-H., Ahmad, Z., Rehman, M. A., & Waseem, M. (2021). Conversion of milk lactose to galactooligosaccharides by enzymes to produce prebiotic enriched cheese. *Future Foods, 4*, 100097.
Rodrigues, L., Paiva, G., Lisboa, H. M., Pasquali, M., Gusmão, R., Duarte, M. E., et al. (2020). Impact of spray drying parameters on lactose-free milk powder properties and composition. *Journal of Agricultural Studies, 8*(3), 32.

San Jose, C., Asp, N. G., Burvall, A., & Dahlqvist, A. (1977). Water sorption in lactose hydrolyzed dry milk. *Journal of Dairy Science, 60*(10), 1539–1543.

Sarkar, S. (2006). Cultured milk products for lactose-intolerant recipients. *Nutrition & Food Science, 36*(5), 357–364.

Schuck, P., Perrone, I. T., Carvalho, A. F., Silveira, A. C. P., Torres, J. K. F., & Bueno, R. G. (2015). Concentração e secagem de produtos lácteos difíceis. *Indústria de Laticínios, 115*, 66–68.

Sharanagouda, B., Lal, H., & Mahajan, I. S. (2022). Study on simultaneous lactose hydrolysis and fermentation in the preparation of lactose hydrolyzed yoghurt and its effect on physico-chemical properties.

Sherwood, S., & Jenkins, D. (2007). US patent US 2007/0178214 A1.

Shrestha, A. K., Howes, T., Adhikari, B. P., & Bhandari, B. R. (2007). Water sorption and glass transition properties of spray dried lactose hydrolysed skim milk powder. *LWT-Food Science and Technology, 40*(9), 1593–1600.

Silanikove, N., Leitner, G., & Merin, U. (2015). The interrelationships between lactose intolerance and the modern dairy industry: Global perspectives in evolutional and historical backgrounds. *Nutrients, 7*(9), 7312–7331.

Singh, P., Rao, P. S., Sharma, V., & Arora, S. (2021). Physico-chemical aspects of lactose hydrolysed milk system along with detection and mitigation of maillard reaction products. *Trends in Food Science & Technology, 107*, 57–67.

Skryplonek, K., Henriques, M., Gomes, D., Viegas, J., Fonseca, C., Pereira, C., & Mituniewicz-Małek, A. (2019). Characteristics of lactose-free frozen yogurt with κ-carrageenan and corn starch as stabilizers. *Journal of Dairy Science, 102*(9), 7838–7848.

Souza, F. P., Balthazar, C. F., Guimaraes, J. T., Pimentel, T. C., Esmerino, E. A., Freitas, M. Q., Raices, R. S. L., Silva, M. C., & Cruz, A. G. (2019). The addition of xyloligoosaccharide in strawberry-flavored whey beverage. *Lebensmittel-Wissenschaft & Technologie, 109*, 118–122.

Torres, J. K. F., Stephani, R., Tavares, G. M., De Carvalho, A. F., Costa, R. G. B., de Almeida, C. E. R., & Perrone, Í. T. (2017). Technological aspects of lactose-hydrolyzed milk powder. *Food Research International, 101*, 45–53.

Tsuchiya, A. C., da Silva, A. D. G. M., Brandt, D., Kalschne, D. L., Drunkler, D. A., & Colla, E. (2017). Lactose-reduced ice cream enriched with whey powder. *Semina: Ciências Agrárias, 38*(2), 749–758.

Vénica, C. I., Bergamini, C. V., Zalazar, C. A., & Perotti, M. C. (2013). Effect of lactose hydrolysis during manufacture and storage of drinkable yogurt.

Walstra, P., Geurts, T. J., Noomen, A., Jellema, A., & van Boekel, M. A. J. S. (1999). *Dairy technology: Principles of Milk properties and processes (Chapter 22)*. Marcel Dekker.

Wan, W., & Xu, B. (2018). Development of an orange juice beverage formulated with oat beta-glucan and whey protein isolate. *Journal of the Science of Food and Agriculture, 98*(12), 4685–4691.

Wolf, M., Gasparin, B. C., & Paulino, A. T. (2018). Hydrolysis of lactose using β-D-galactosidase immobilized in a modified Arabic gum-based hydrogel for the production of lactose-free/low-lactose milk. *International Journal of Biological Macromolecules, 115*, 157–164.

Zagorska, J., Ciprovica, I., Straumite, E., & Majore, K. (2020). Acceptance of low-sugar yoghurt among Latvian teenagers. *Agronomy Research, 18*(S3), 1897–1905.

Chapter 7
Non-dairy Dietary Approach for Lactose Intolerant

Nikheel Rathod, Arvind Soni, Priyanka Meena, and Heena Sharma

7.1 Introduction

Lactose (milk sugar) is the major carbohydrate present in mammalian milk and is a disaccharide consisting of -D-glucose (or -D-glucose) and -D-galactose (Harrington & Mayberry, 2008). Lactose is normally found in dairy products in either of crystalline forms or amorphous form. Lactase is an enzyme produced in the mucosa of the intestine that catalyzes the breakdown of lactose. Lactose intolerance (LI) is a metabolic condition characterized by an inadequate presence of the lactase enzyme (β-galactosidase [β-gal]), which is responsible for breaking down lactose into simpler sugars called glucose and galactose (Suchy et al., 2010). Developmental Lactase deficiency seen in premature infants born at 28–37 weeks of gestation (Mobassaleh et al., 1985). For those who are intolerant to lactose and must abstain from dairy products, there are several non-dairy dietary strategies that can guarantee sufficient nourishment. For those who are lactose intolerant and need to avoid dairy products, there are a number of non-dairy dietary strategies that can ensure enough nutrition. A range of plant sources, such as seeds, nuts, legumes, cereals, and pseudo-cereal plants, are used to make plant-based substitutes for dairy milk (Mäkinen et al., 2016). Lactose intolerant people can maintain a healthy and balanced diet by utilizing non-dairy substitutes such soy milk, almond milk, rice milk, oat milk, and coconut milk, as well as making sure they get enough calcium and other important

N. Rathod (✉)
PG Institute of Post-Harvest Technology and Management (Dr. Balasaheb Sawant Konkan Krishi Vidyapeeth), Raigad, Maharashtra, India

A. Soni · P. Meena
ICAR-Central Sheep & Wool Research Institute, Avikanagar, Rajasthan, India

H. Sharma
Dairy Technology Division, ICAR-National Dairy Research Institute, Karnal, Haryana, India

P. S. Rao et al. (eds.), *Lactose Hydrolysis in Dairy Products*,
https://doi.org/10.1007/978-3-031-78207-7_7

nutrients. Developments in the food and beverage manufacturing sector have made it easier to produce lactose-free milk, which removes the lactose from animal milk and makes it easier for people who are lactose intolerant to digest.

Therefore, there is a great chance to look into plant-based substitute's creation of milk-like beverages. Additionally, because they contain nutrients that improve health, like dietary fiber, minerals, vitamins, and antioxidants, plant sources like cereals and legumes have recently come to be recognized as functional foods and nutraceuticals (Moore et al., 2024). We have tried to shed light on the current wave of non-dairy options available to those who are lactose intolerant through this chapter. The substitutes discussed come from seeds, nuts, legumes, cereals, and pseudocereals. Furthermore, the technological challenges in their usage and consumer acceptability, their comparison with dairy nutrition.

7.2 Dairy-Free Plant-Based Beverages and Products

Plant-based milk products offer a valuable mineral source and additionally contain non-allergenic proteins and essential fatty acids, making them a viable substitute for animal milk, as observed (Robayo-Torres & Nichols, 2007). Recently, these products have been marketed as functional and nutraceutical foods due to their bioactive compounds, including antioxidants, dietary fibers, vitamins, and minerals, which contribute to their health benefits (Olías et al., 2023). The alternatives to plant-based milk predominantly originate from legumes and nuts, as these ingredients possess unique properties that facilitate their combination and production, resulting in dairy-free, cost-effective, nutritious, flavorful, and, most importantly, health-conscious alternatives, as demonstrated by Rasinperä et al., 2005.The classification of lactose free milk alternatives has been proposed (Paul et al., 2019; Sethi et al., 2016) providing relief from lactose intolerance (Fig. 7.1).

7.2.1 Oat Milk

Oat milk has recently gained prominence in the market due to its potential therapeutic advantages. Oats have garnered significant attention for their rich content of dietary fibers, phytochemicals, and high nutritional value. Oats offer various health benefits, including cholesterol-lowering and potential anticancer properties. These benefits are attributed to the presence of dietary fibers like β-glucan, functional proteins, lipids, starch components, and phytochemicals found within the oat grain. As a result, oats have become a promising raw material for producing functional plant-based milk. Notably, oats are an excellent source of high-quality protein with a well-balanced amino acid profile. The interest in oats is primarily driven by the presence of the functionally active component β-glucan, which exhibits nutraceutical properties. β-glucan, a soluble fiber, has the ability to increase solution viscosity,

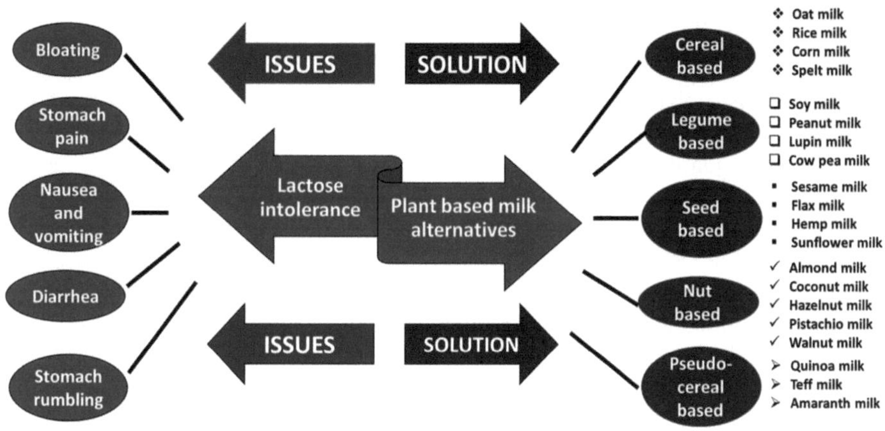

Fig. 7.1 Classification of lactose free milk alternative and their health benefits

delay gastric emptying time, and prolong gastrointestinal transit time, leading to reduced blood glucose levels (Welch, 1995). Moreover, oats are rich in antioxidants, polyphenols, and nutrients, comprising approximately 60% starch, 11–15% total protein, 5–9% lipids, 2.3–8.5% dietary fiber, and 0.54% calcium (Rasane et al., 2015). Oats have been extensively studied for their utility in gluten-free or celiac diets. To diversify oat consumption, efforts have been made to develop oat-based beverages or oat milk (Deswal et al., 2014). Research has demonstrated that oat beverages inherently retain their cholesterol-lowering properties (Önning et al., 1999). Zhang et al. (2007) conducted treatments on oat-based beverages with phytase to enhance nutritional value by releasing inorganic phosphate from phytic acid. They also made efforts to replenish nutrients inevitably lost during processing operations through enrichment. These oat milk products are conveniently available in various sizes in tetra packs and undergo UHT (Ultra High Temperature) treatment.

Recently Benmeziane, & Belleili, (2023) reported development of novel yoghurt using fermented oat milk. The findings indicated that fermented oat milk had an extended shelf life of 7 days due to its increased resistance to deterioration caused by the growth of mold and yeast. When compared to cow milk, fermented oat milk showed higher sensory scores during storage. Considering the increased demand for plant-based milk, due to increased lactose intolerance amongst the populations. Zhou et al., (2023) compared different oat cultivars from Australia and China for extraction of milk. The findings suggested, Chinese cultivar (Chinese Bayou 01 [ZBY01]) produced milk exhibited higher stability which is comparable with treated milk samples. The milk form ZBY01 exhibited lowest separation rate, highest consistency index, smallest particle size (0.48 μm) and lowest sedimentation was observed in comparison with treated milk. Considering the nutritional advantage of oat milk, it was found to contain higher proportion of carbohydrates, vitamin B1, vitamin E, iron and phosphorous over cow milk. Additionally, add dietary fibers, minerals, tryptophan and phytosterols over the cow milk. Also, the inclusion of oat milk has been reported to prevent diabetes and cardiovascular diseases (Yu et al., 2023).

7.2.2 Soy Milk

The soybean is globally recognized as one of the most extensively cultivated and popularly consumed legumes. Soybeans boast an impressive nutritional profile, featuring a high-quality protein that encompasses all the essential amino acids. Additionally, they offer substantial quantities of essential minerals, including calcium, copper, iron, magnesium, phosphorus, potassium, and zinc, while also being highly digestible (Rasinperä et al., 2005). Soya milk, rich in isoflavones, offers numerous health benefits, including cholesterol reduction, menopause relief, osteoporosis prevention, and cancer risk reduction. Its high intake also protects cells and DNA from oxidation, making it a valuable source of antioxidants. Soy foods are rich in phytochemicals, such as phytosterols, which have cholesterol-lowering properties (Fukui et al., 2002). Dry soybean consists of 35% carbohydrates, 40% protein, 20% oil, and 5% ash (Liu, 2004).

Soy milk was the first plant-based milk, providing nutrients to populations with inadequate milk supply and lactose intolerant populations. Isoflavones serve as the bioactive compounds responsible for their safeguarding effect against cancer, cardiovascular diseases, and osteoporosis (Omoni & Aluko, 2005). Traditional production methods have limited shelf life and beany flavor. Modern production uses advanced technologies to maximize nutritional value, shelf life, and convenience. Soy milk is offered in different variants based on solids concentration, including light, dairy-like, and rich options. It also comes in various formulations like sweetened, original, and flavored, and can be found in three fortification categories: regular, enriched, and blended (Liu, 2004).

Maftei et al. (2023) made an attempt to create a fermented beverage made of soy milk in light of the growing trend of people consuming fermented products. By increasing the viability of Bifidobacteria and their survival in the human gut, the addition of buckthorn powder affected the beverage's quality and demonstrated the probiotic effect of lactose-free soy milk. Cui et al. (2021) assessed the impact of various combinations of microbial starter cultures on the quality characteristics of soy-milk yoghurt. In comparison to cow milk yoghurt, the formulated yoghurt showed a higher count of *S. thermophilus*, compact network formation with firm structure due to excellent protein coagulation, and stable pH during storage. Yogurt made from cow's milk, however, had higher levels of the probiotic strain Bifidobacterium animalis subsp. Lactis BB-12 growth. Proposing soy milk as a viable substitute for probitoc enrichment in individuals who are lactose intolerant (Cui et al., 2021). Additionally, it has been reported that soy milk and its derivatives work well as nutraceuticals to prevent hyperglycemia (De et al., 2022). Okhara and yoghurt, two products made from soy milk, decreased lipid parameters and the spike in blood sugar. The authors proposed that soy milk and related products' high fiber content had a glycaemic control and lipid-lowering effect.

7.2.3 Peanut Milk

Peanuts are a commonly consumed oilseed among individuals with limited financial resources, those who have milk allergies, nutrient-deficient children, and certain vegetarians (Sethi et al., 2016). They boast a substantial nutritional profile, containing significant amounts of proteins (23.68%), carbohydrates (21.5%), fats (49.6%), as well as vitamins, minerals, phytochemicals (PSs), antioxidants (AOXs), and volatile compounds (Settaluri et al., 2012). The functional attributes of peanuts primarily stem from their content of phenolic compounds, renowned for their antioxidant properties and their ability to offer protection against diseases characterized by oxidative damage, such as coronary heart disease, stroke, and various forms of cancer. Furthermore, peanuts have been associated with improvements in blood lipid and sugar levels, contributing to overall longevity (Wien et al., 2014). When combined with cowpea milk to create a fermented beverage, peanut milk demonstrated greater antioxidant activity than fermented cow milk (Chawafambira et al., 2022). With a higher number of viable cells, the peanut milk-based matrix promoted the growth of *Lacticaseibacillus rhamnosus*. According to Chawaffambira et al. (2022) the amino acid composition of the peanut milk and cow pea milk samples was similar to that of fermented milk samples, and the minerals potassium, magnesium, and iron were significantly higher than those found in milk. Peanuts share many volatile compounds with soybeans (Leu, 1974). They are recognized as a healthful food due to their various bioactive components known for their disease-preventive properties. In terms of composition, peanuts consist of approximately 21.5% carbohydrates, 49.6% fats, 23.68% proteins, and 8% crude fiber (Settaluri et al., 2012). It was discovered that the development of peanut allergen is inhibited by fermented peanut milk made with *Lactiplantibacillus plantarum* Kinko-SU4 (Yamamoto et al., 2022). It was discovered that fermentation enhanced the fermented peanut milk's ability to scavenge free radicals, which resulted in enhanced anti-oxidant and anti-glycation qualities. The process involves immersing and grinding whole peanuts with water to create a thick mixture, which is then filtered. This milk-like substance is subsequently homogenized and pasteurized, following a similar procedure to fresh milk. Additionally, it may be enriched with vitamins and minerals and occasionally flavored (Chan & Beuchat 1992). However, Sakthi et al. (2020) assessed the effects of several pre-treatments (soaking, blanching, roasting, and germination) on the quality of peanut milk. It was discovered that the blanched peanut milk produced milk with a high level of sensory acceptability. According to recent developments, the use of a microfluidization system at 120 MPa produced peanut milk with a high level of physical stability, smaller particle size, and improved dispersibility. In addition, the extraction methods improved the peanut milk's flavor profile (Dai et al., 2022). Recommending as a successful substitute method for peanut milk extraction.

7.2.4 Almond Milk

Almond (*Prunusdulcis*) is one of most popular and widely consumed of nuts (Nande et al., 2008). High consumption rate among the humans is probably due to its well-known medicinal properties such as healthy lipid profile, immune enhancing and antioxidant effects. The high demand for almonds can be attributed to their well-established health benefits, including the promotion of a healthy lipid profile, immune system enhancement, and antioxidant effects. Almonds are a concentrated source of nutrients, notably containing a significant amount of α-tocopherol (36.4%), which plays a crucial role in combating free radicals and preventing oxidative stress (Niki et al., 1989).

Considering the demand for alternate plant base milk, almond milk has gained importance. Based on the vitamin composition of almond milk, it has highest proportion of all the vitamins in comparison to other plant sources of milk, making almond milk an important plant source of milk (Romulo, 2022). In Devnani et al. (2020) assessment, the effects of thermal treatment (45–95 °C) on almond milk were examined with regard to both skim and full fat varieties. Heat above 55 °C reduced the stability of almond milk, and gels with a high water-holding capacity were formed between 85 and 95 °C, indicating that almond milk proteins could be used as an gelling ingredient in vegan products. Considering the disadvantages of the traditional almond milk extraction technologies and the nutritional benefits of almond milk, several studies have focused on the pretreatment with several novel technologies for enhanced recovery and quality almond milk (Manzoor et al., 2017; Vanga & Raghavan, 2018).

The application of thermosonication for pasteurizing almond milk was assessed by Mattison et al. (2020). By reducing the particle size distribution and increasing microbial (total plate count, yeast and mold count), enzymatic (lipoxygenase and peroxidase) inactivation, thermosonication improved the quality. Thermosonication enhanced the almond milk's bioactive composition (phenolic, flavonoid, flavonol, and condensed tannin), in addition to its physical and microbiological quality. Almond milk's antioxidant activity is enhanced along with a rise in bioactive components. While the impact of ultrasound on the protein content of almond milk was assessed (Vanga & Raghavan, 2018). The use of ultrasound led to a restructuring of the protein structure (loss of the α-helix structure), and treatment duration was found to have an impact.

Recent research has concentrated on the development of fermented products using almond milk, given its high nutritional value (Topcuoglu, & Yilmaz-Ersan, 2020; Zahrani, & Shori, 2023; Yilmaz-Ersan, & Topcuoglu, 2022). Since almond milk contains fermentable sugars that support metabolic activity, it has been found to support the viability of probiotic microorganisms (Topcuoglu, & Yilmaz-Ersan, 2020). The total phenolic and flavonoid content of almond milk has been shown to increase significantly with fermentation, and the composition was found to be influenced by the use of different microbial strains, with *L. rhamnosus* having the highest concentration of phenolics and flavonoid. Almond milk is a viable option for delivering probiotic strains to individuals who are lactose intolerant. Moreover, almonds are rich in essential minerals such as magnesium (19.5%), copper (16.0%), and phosphorus (13.4%). They are also notably high in dietary fiber, accounting for approximately 13.2% of their composition. Almonds are protein-rich, comprising around 25% protein content, with a particularly high content of arginine and the majority of the protein is present in the form of amandin (Sathe et al., 2002).

While almonds have a relatively high fat content (49.4% of weight), the type of fat they contain is predominantly monounsaturated fat (MUFA), making up 67% of their fat composition. This high MUFA content is known to have potential benefits for heart health. Additionally, almonds contain polyphenols and phytosterols, such as β-sitosterol, stigmasterol, campesterol, sitostanol, and campestanol, which confer cardioprotective properties (Berryman et al., 2011). Being a rich source of calcium and fat and low in calorie count, it is nutritionally better than other plant-based milks.

7.2.5 Coconut Milk

Coconut milk is a milky fluid extracted from fresh coconut endosperm through either manual or mechanical methods, with or without the addition of water (De Leon et al., 2005; Tulashie et al., 2022). Coconut is a highly nutritious product and serves as a rich source of dietary fiber. Coconut milk is abundant in essential vitamins and minerals, including iron, calcium, potassium, magnesium, and zinc. It also contains a significant amount of vitamin C and E (Seow & Gwee, 1997). The consumption of coconut milk has been associated with various health benefits, such as its potential anti-carcinogenic, antimicrobial, antibacterial, and antiviral properties. It contains lauric acid, a saturated fat also found in mother's milk, which is believed to support brain development (Belewu & Belewu, 2007; Karunasiri et al., 2020; Bharti et al., 2021; Chen et al., 2024a). Coconut milk's immuno-protective effects are attributed to lauric acid, known for its antimicrobial properties that inhibit the growth of specific microorganisms like Mycobacterium tuberculosis and Staphylococcus aureus, among others (Barr, 2013). Furthermore, coconut milk offers additional advantages, including aiding in digestion, nourishing the skin, and possessing cooling properties. It boasts a creamy texture and a subtly sweet coconut flavor. In terms of nutritional comparison with cow's milk, coconut milk contains only a third of the calories, half the fat, and significantly lower levels of protein and carbohydrates (Bailey et al., 2013; Chen et al., 2024a).

7.2.6 Sesame Milk

Sesame holds a significant position among the world's oilseed crops. It stands out as a source of high-quality protein, boasting a unique blend of amino acids. Sesame seeds contain notable quantities of lignans, including sesamin, sesamolin, sesaminol, and others, renowned for their functional properties (Chen et al., 2024b). These sesame lignans are acknowledged for their nutraceutical attributes, such as antioxidative, hypocholesterolemic, anticarcinogenic, antitumor, and antivirus properties (Namiki, 2007; Aydar et al., 2020). These factors limit their use in the production of plant-based milks. Hence, there's a need to modify the functionality of sesame protein before incorporating it into plant-based milk (Quasem et al., 2009). Studies have revealed that techniques like steaming and alkali soaking significantly increase the total solids yield (Prakash & Narasinga Rao, 1986; Aydar et al., 2020), while pasteurization improves dispersion stability. Roasting and alkali soaking have been observed to enhance overall acceptability and flavor by reducing bitterness and chalkiness in sesame milk. Sesame milk consumption offers an alternative to the limitations associated with soy milk, such as the presence of flatulence-inducing factors, allergies to soy proteins, and undesirable beany or off flavors (Zahra et al., 2014; Ahmadian-Kouchaksaraei et al., 2014).

7.2.7 Hemp Milk

Hemp milk is made from the seeds of the hemp plant (*Cannabis sativa*) (Segal & Blair, 1961). Hemp milk is rich in protein and vitamins such as A and E, alternative to milk for individuals with lactose intolerance (Sevillano Pires et al., 2022). Hemp milk provides approximately 30% of the recommended daily intake of vitamin D and a significant amount of calcium (Sevillano Pires et al., 2022). Furthermore, this hemp-based beverage serves as an excellent source of essential minerals, including magnesium, potassium, and iron. Since hemp milk is derived from hemp seeds, its potential positive impacts on human health hinge on the levels of linolenic acid (C18:2 ω-6) and α-linolenic acid (C18:3 ω-3). Its high content of polyunsaturated fatty acids (PUFAs) offers benefits for various populations by reducing the accumulation of harmful cholesterol, regulating blood pressure, stimulating neurons, and enhancing mental acuity (Besir et al., 2022; Curl et al., 2020; Rizzo et al., 2023; Al-Khazaleh et al., 2024). It is particularly effective as a substitute for lighter milk options, such as skimmed milk. Its robust flavor is well-suited for savory dishes, and its protein content makes it a valuable ingredient in baking.

7.2.8 Rice Milk

Rice milk, a well-known milk substitute, has become more popular recently since it is suitable for people who are lactose intolerant. Although rice milk is low in protein, it is high in critical minerals including selenium and manganese as well as vitamin B (Heyman, 2006; da Silva et al., 2023). It takes grinding rice and combining it with water to make rice milk. The process of making rice milk involves milling rice and mixing it with water, with the option of fortifying it with added calcium and vitamin D. According to da Silva et al. (2023), rice milk is thought to have a high antioxidant capacity because of its rich supply of phenolic compounds. Moreover, it is advised to hydrate the grains at 25 °C for 6 h in order to extract rice milk with a high nutritional content, and slow pasteurization is advised for sterilization. It has recently been assessed how hydrogen-rich water affects the nutritional and sensory qualities of rice milk (Zor et al., 2024). Use of hydrogen rich water was found to improve the nutritional (essential minerals: sodium, magnesium, potassium, calcium and selenium, sugars, amino acids: isoleucine, leucine and methionine) and sensory quality (aroma profile) of rice milk. Considering the lower protein content of rice milk, inclusion of protein hydrolysate along with ultrasonication treatment improved bioactivity and in-vitro intestinal absorption (Ngamsuk et al., 2020). Rice milk has been used in the formulation of various value-added products (such as probiotics, ice cream, beverages, etc.) based on its acceptability (Rasika et al., 2021; Mygdalia et al., 2023).

Thinking about the shift in diet toward non-dairy products in order to create cheese-like food, Nakamura et al. (2016) fermented rice milk with lactic acid bacteria while also utilizing a number of coagulation agents. The rice milk-based cheese-like product that was created showed strong consumer acceptability. A fermented probiotic drink made with brown rice milk was made (Kittibunchakul et al., 2021). The rice milk increased the proliferation of probiotic *Lactobacillus pentosus* 9D3 above the required limit. The probiotic beverage had significant antioxidant activity and a high total phenolic content, both of which were beneficial to health. Furthermore, the beverage's ability to inhibit certain enzymes (lipase: 22.62%, α-amylase: 61.97%, α-glucosidase: 64.50%) makes it a potential treatment for a few chronic illnesses.

7.2.9 Cashew Milk

Cashew milk offers an excellent alternative to dairy for those with lactose intolerance and dairy allergies. It serves as a lower-calorie alternative that delivers a richer consistency than standard soy, rice, and almond-based offerings. Cashew milk is produced by blending cashew nuts with water (Manzoor et al., 2017). Studies found that cashew milk contains high levels of energy, protein (23%), carbohydrate (29.30%), essential fatty acids (44%), unsaturated fats (82%), and minerals (4.75 mg calcium, 7.15 mg potassium, 3.00 mg iron, and 2.00 mg magnesium and phosphorus; (Bruno et al., 2019). Cashew milk has gained popularity due to its potential health benefits, such as antioxidant properties, diabetes management, bone health maintenance, cholesterol reduction, and hypertension prevention (Mattison et al., 2020; Shori et al., 2022).

Additionally, cashew milk is often fortified with essential nutrients like calcium and vitamin D, helping to maintain strong bones and teeth. It is generally lower in calories than cow's milk and is a good source of antioxidants like vitamin E. The creamy texture of cashew milk makes it an excellent option for adding to coffee, cereal, and smoothies, making it easier for those with lactose intolerance to enjoy their favorite foods without digestive discomfort (Grasso et al., 2020). However, it's essential to choose unsweetened varieties or make homemade cashew milk to control sugar content and select brands with minimal additives for the healthiest option. Cashew milk is nutrient dense as compared to soy and dairy milks with less calories. Cashew milk has been shown to support cardiovascular health and mitigate deficiencies in trace minerals like zinc and iron, as demonstrated in a study by Manzoor et al., 2017.

7.2.10 Macadamia Milk

An increasing number of individuals are choosing non-dairy milk options, and one such option is macadamia milk, which surprisingly is quite simple to prepare and offers a delightful, delicate flavor. It can be enjoyed by itself, added to smoothies, or poured over cereal (Camacho-Teodocio et al., 2024). Macadamia milk is a fresh addition to the world of non-dairy milk alternatives, crafted from macadamia nuts. To make it, macadamia nuts are roasted, boiled or soaked in water and then blended to create creamy milk (Kim et al., 2022). This milk can be enhanced with various flavors like vanilla or chocolate. Notably, macadamia milk provides a good dose of both protein and calcium while remaining low in calories (Borin et al., 2022). Additionally, it stands out as an environmentally friendly choice because it is derived from a renewable resource. Macadamia milk boasts significant levels of unsaturated fatty acids, establishing it as a noteworthy nutritional resource (Kim et al., 2022). Abundant in calcium, vitamin D, and vitamin B12, this milk alternative offers a dense texture, rendering it an excellent pick for enhancing coffee beverages (Sethi et al., 2016). As substitutes to dairy milk, other plant-based milks are also developed, including hazelnut, quinoa, cow pea, sunflower, hemp and melon seed milk (Bastıoğlu et al., 2016).

7.3 Technological Challenges and Consumer Acceptability of Plant-Based Milk

Producing non-dairy alternatives with lactose intolerance involves several technological challenges with perspective of its acceptability among consumers:

(a) **Similarity in taste and texture:** One of the primary challenges is replicating the taste and texture of dairy products. Dairy alternatives must mimic the creamy texture and pleasant flavor of milk, cheese, yogurt, etc., to appeal to consumers. Despite being designed to have a mouthfeel and color similar to animal milk, many plant-based alternatives have undesirable qualities. Such as some have beany or painty off-flavours due to lipoxygenase activity (Vanga &

Raghavan, 2018), while others have a chalky mouthfeel caused by large insoluble particles (Sakthi et al., 2020), and still others have a darker appearance than cow's milk (Pointke et al., 2022). The acceptability of peanut milk has been observed to rely on the color, texture, lack of peanut flavor, and likeness to cow's milk (Diarra et al., 2005). It was discovered that the volatile compounds in lactose-free plant-based milk affected its sensory quality attributes (Pointke et al., 2022). It was discovered that the plant source, such as cereal and nutty flavor for milk made from cereal and nut bases, had an effect on the sensory character's sensory quality (Vaikma et al., 2021).

(b) **Nutritional value:** Non-dairy alternatives may be poor source calcium, vitamin D, and other nutrients than dairy products and they need to be requiring fortification to match the nutritional profile of dairy products (McClements, 2020). Ensuring adequate fortification that maintains bioavailability and consumer acceptance is crucial. The main reasons stated by consumers for the use of plant-based milks are digestibility and lactose intolerance, or malabsorption of the lactase enzyme (Haas et al., 2019). Another frequent justification for preferring dairy substitutes is perceived healthfulness (Boaitey et al., 2020).

(c) **Quality, Consistency and supply chain:** Ensuring high-quality ingredients for non-dairy products with quality standards is crucial, especially if they are plant-based or derived from specific sources like nuts or soy. Variations in quality can affect the taste, texture, and nutritional content of the final product. Managing supply chains to ensure consistent quality and availability of ingredients is essential.

(d) **Operational and fabrication methodologies:** The manufacturing procedures for non-dairy alternatives may differ significantly from those used in dairy processing. Developing efficient and scalable techniques for extracting, processing, and manufacturing non-dairy ingredients while preserving nutritional value and taste is a technological hurdle.

(e) **Mitigation of individual allergy:** Variety of non-dairy alternatives such as nuts or soy, may produce individual allergy. Developing methods to prevent cross-contamination and ensuring allergen-free products are crucial for consumer safety.

(f) **Shelf Stability of the product:** Maintaining stability and a prolonged shelf life for non-dairy products may be a challenging task to avoid spoilage, or off-flavors. Developing stabilizers, emulsifiers, and preservatives that are suitable for plant-based ingredients can be difficult.

(g) **Economic efficiency:** Ensuring cost efficiency in production, while maintaining quality standards and meeting nutritional requirements, is crucial for enhancing accessibility of non-dairy alternatives to a wider consumer base.

(h) **Public awareness:** Non-dairy products as alternative to lactose intolerance often requires educating consumers about their benefits, nutritional content, and how they can be consumed with their diets. Effective communication is essential for acceptance and adoption.

(i) **Government and legal compliance:** Meeting regulatory standards and ensuring compliance with labeling requirements for allergens, nutritional claims, and ingredient declarations is crucial for market entry and consumer trust.

Addressing these technological challenges requires ongoing innovation, collaboration between food scientists, engineers, and nutrition experts, and a commitment to meeting consumer demand for diverse dietary choices.

7.4 Comparison of Nutrients with Dairy Products

Plant-based milks often undergo fortification to align their micronutrient composition more closely with that of dairy milk. However, significant variations persist, particularly in terms of protein, zinc, and potassium, which can differ based on the choice of base ingredient and the specific product. Additionally, certain plant-based milks incorporate extra sugar to enhance their flavor. The nutritional quality and flavor of plant-based milks exhibit considerable variation. The overall content of protein, fat, and nutrients is contingent upon the primary ingredient. Fat levels differ in both dairy and plant-based milks. Notably, several plant-based milks, such as almond and cashew milk, have substantially lower protein and other essential nutrients for growth and well-being, including zinc, potassium, and magnesium. Additionally, plant-based milks often incorporate added sugars, flavors, or ingredients to enhance taste and texture, thereby influencing their overall health profile (Table 7.1).

7.5 Significant Health Implications

Consumers tend to prefer plant-based milk alternatives, such as soy milk, over dairy alternatives (McCarthy et al., 2017). Plant milk substitutes are often perceived as healthy, possibly due to negative perceptions about cow's milk nutrition and the health claims associated with soy (Patisaul and Jefferson, 2010). It depends greatly on the raw material, the processing, the fortification, and other ingredients like sweeteners or oils for their nutritional properties. Some plant derived components have favorable health effects, that may be present in the beverages produced from that raw material. Plant-based diets with little or no animal products are related with a decreased incidence of diabetes, CVD, certain cancers, and healthier weights (BMI) in adults (Dinu et al., 2017). Substituting skim cow's milk with oat or soy alternatives has been documented to reduce the levels of plasma cholesterol and low-density lipoprotein (LDL) in healthy individuals after a 4-week consumption period (Önning et al., 1998). Some plant-based milk contains bioactive compounds, such as isoflavones and phytosterols in soy milk, or nutritional factors such as phytates in oat milk. (Yuan et al., 2022) which has been linked to potential benefits in

Table 7.1 General composition of plant-based milk substitute and Cow milk

Type of milk	Energy (kcal)	Protein (g)	Total Lipids (g)	Carbohydrate (g)	Total Fibers (g)	Total Sugar (g)	Calcium (mg)	References
Soybean milk	46	3.33	1.88	3.75	0.8	2.50	188	USDA (2019)
Almond milk	42	0.42	1.04	7.92	0.4	7.5	188	USDA (2019)
Rice milk	47	0.1	1	9.5	0	3.3	120	Alpro (2019)
Oat milk	41.7	1.67	0.63	7.5	0.9	2.09	7.5	Elmhurst, 1925 (2019)
Hazelnut milk	121	0.4	1.60	3.1	0.3	3.1	120	Alpro (2019)
Cashew nut milk	66	1.9	4.65	4.44	0.8	1.27	8	USDA (2019)
Sesame seed milk	–	2.97	6.43	4.04	–	–	–	Afaneh et al. (2011)
Chickpea milk	–	1.21	0.34	2.32	0.78	–	–	Wang et al. (2008)
Quinoa milk	121	0.5	1.2	3.5	–	0.3	–	ECOMIL (2019)
Sunflower seed milk	21.08	0.78	2	0	1.2	–	0.16	Blum et al. (2016)
Cow milk	67	3.33	3.33	5.42	0	5	129	USDA (2019)

averting conditions such as cardiovascular ailments, prostate cancer, and osteoporosis (Patisaul and Jefferson, 2010). Zhang et al. (2007) achieved the most significant enhancement in iron bioavailability in fortified oat milk by employing citric acid in conjunction with exogenous phytase. Almond milk is made from almonds, which are abundant in minerals and a good source of protein and vitamin E (Chen et al., 2006). Despite being high in nutrients, almond milk has certain undesirable sensory qualities (Tangyu et al., 2019). According to Chalupa-Krebzdak et al. (2018), oat milk is another PBA that offers a number of health advantages, including fiber, phytochemicals, and its total nutritional worth. In addition, oats can reduce insulin and glucose responses as well as blood cholesterol levels (Deswal et al., 2014). Emerging studies on the function of the microbiota may give insight on the ability of plant-based diets to modify inflammatory indicators and alter bone remodeling pathways, hence maintaining bone health in the long run (Tomova et al., 2019). A recent review of the research indicated that plant-based diets are not hazardous to adult bone health as long as calcium and vitamin D intake is adequate (Hsu, 2020).

7.6 Conclusions

In conclusion, the non-dairy dietary approach for individuals with lactose intolerance offers a viable and beneficial alternative to traditional dairy-based diets. As we have explored, dairy-free plant-based beverages and products have emerged as a versatile and accessible option for those who cannot tolerate lactose. Despite technological challenges and varying consumer acceptability, these non-dairy alternatives have evolved to meet diverse tastes and preferences. Furthermore, to enhance consumer acceptance through technological advancements, the development of plant-based milk alternatives will remain a prominent focus in the emerging field of food science and technology. Despite the challenges related to taste and nutrition that plant-based products need to address to establish themselves as competitive alternatives to cow's milk, their environmental sustainability plays a pivotal role in their increasing popularity. Many people's intolerance to the lactose in cow's milk, dairy allergies, consumer concerns about hormones and antibiotics in cow's milk, the high cholesterol content of cow's milk, concerns about animal welfare, growing interest in vegetarian and vegan diets, and the potential health benefits of a plant-based diet are additional factors driving the growth of this market (Tangyu et al., 2019).

References

Afaneh, I., Abu-Alruz, K., Quasem, J. M., Sundookah, A., Abbadi, J., Alloussi, S., & Sawalha, S. (2011). Effect of critical processing variables on sesame milk quality. *American Journal of Agricultural and Biological Science.*

Ahmadian-Kouchaksaraei, Z., Varidi, M., Varidi, M. J., & Pourazarang, H. (2014). Influence of processing conditions on the physicochemical and sensory properties of sesame milk: A novel nutritional beverage. *LWT-Food science and Technology, 57*(1), 299–305.

Al-Khazaleh, A. K., Zhou, X., Bhuyan, D. J., Münch, G. W., Al-Dalabeeh, E. A., Jaye, K., & Chang, D. (2024). The Neurotherapeutic Arsenal in Cannabis sativa: Insights into Anti-Neuroinflammatory and Neuroprotective Activity and Potential Entourage Effects. *Molecules, 29*(2), 410.

Alpro. Rice original. (2019). Retrieved from https://www.alpro.com/uk/products/drinks/rice-drinks/rice-original/

Aydar, E. F., Tutuncu, S., & Ozcelik, B. (2020). Plant-based milk substitutes: Bioactive compounds, conventional and novel processes, bioavailability studies, and health effects. *Journal of Functional Foods, 70*, 103975.

Bailey, R. K., Fileti, C. P., Keith, J., Tropez-Sims, S., Price, W., & Allison-Ottey, S. D. (2013). Lactose intolerance and health disparities among African Americans and Hispanic Americans: An updated consensus statement. *Journal of the National Medical Association, 105*(2), 112–127.

Barr, S. I. (2013). Perceived lactose intolerance in adult Canadians: A national survey. *Applied Physiology, Nutrition, and Metabolism, 38*(8), 830–835.

Bastıoğlu, A. Z., Tomruk, D., Koç, M., & Ertekin, F. K. (2016). Spray dried melon seed milk powder: Physical, rheological and sensory properties. *Journal of Food Science and Technology, 53*(5), 2396–2404.

Belewu, M. A., & Belewu, K. Y. (2007). Comparative physicochemical evaluation of tiger nut, soybean and coconut milk sources. *International Journal of Agric and Biology, 9*(5), 785–787.

Benmeziane, F., & Belleili, I. (2023). An innovative approach: Formulation and evaluation of fermented oat milk using native yoghurt cultures. *Measurement: Food, 12*, 100113.

Berryman, C. E., Preston, A. G., Karmally, W., Deckelbaum, R. J., & Kris-Etherton, P. M. (2011). Effects of almond consumption on the reduction of LDL cholesterol: A discussion of potential mechanisms and future research directions. *Nutrition Reviews, 69*, 171–185.

Besir, A., Awad, N., & Mortaş, M. (2022). A plant-based milk type: hemp seed milk bitkisel sütlerden biri: kenevir tohumu sütü. *Akademik Gida, 20*(2).

Bharti, B. K., Badshah, J., & Beniwal, B. S. (2021). A review on comparison between bovine milk and plant based coconut milk. *Journal of Pharmaceutical Innovation, 10*(3), 374–378.

Blum, J. E. S., Ramoni, E. O., & Balbi, M. E. (2016). Elaboração de extratohidrossolúvel (leite) a partir de semente de girassolgerminada (Helianthus annus L., Asteraceae) e avaliação de suacomposiçãonutricional. *VisãoAcadêmica, 17*, 81–95.

Boaitey, A., & Minegishi, K. (2020). Determinants of household choice of dairy and plant-based milk alternatives: Evidence from a field survey. *Journal of Food Products Marketing, 26*(9), 639–653.

Borin, J. F., Knight, J., Holmes, R. P., Joshi, S., Goldfarb, D. S., & Loeb, S. (2022). Plant-based milk alternatives and risk factors for kidney stones and chronic kidney disease. *Journal of Renal Nutrition, 32*(3), 363–365.

Bruno, L. M., Lima, J. R., Wurlitzer, N. J., & Rodrigues, T. C. (2019). Non-dairy cashew nut milk as a matrix to deliver probiotic bacteria. *Food Science and Technology, 40*, 604–607.

Camacho-Teodocio, J. D., Gallardo-Velázquez, T., Osorio-Revilla, G., Castañeda-Pérez, E., Velázquez-Contreras, C., Cornejo-Mazón, M., & Hernández-Martínez, D. M. (2024). Macadamia (Macadamia integrifolia) Nut-Based Beverage: Physicochemical Stability and Nutritional and Antioxidant Properties. *Beverages, 10*(3), 58.

Chalupa-Krebzdak, S., Long, C. J., & Bohrer, B. M. (2018). Nutrient density and nutritional value of milk and plant-based milk alternatives. *International Dairy Journal, 87*, 84–92.

Chan, L., & Beuchat, L. (1992). Chemical, physical and sensory characteristics of peanut milk as affected by processing conditions. *Journal of Food Science, 57*, 401–405.

Chawafambira, A., Jombo, T. Z., & Mkungunugwa, T. (2022). Effect of Lacticaseibacillus rhamnosus yoba fermentation on physicochemical properties, amino acids, and antioxidant activity of cowpea-peanut milk. *Journal of Food Quality, 2022*(1), 3192061.

Chen, C. Y., Lapsley, K., & Blumberg, J. (2006). A nutrition and health perspective on almonds. *Journal of Science and Food Agriculture, 86*, 2245–2250.

Chen, Y., Li, T., Jiang, L., Huang, Z., Zhang, W., & Luo, Y. (2024a). The composition, extraction, functional property, quality, and health benefits of coconut protein: A review. *International Journal of Biological Macromolecules*, 135905.

Chen, T., Wang, B., Yang, Y., Han, B., & Jiang, H. (2024b). Sesame use in Turpan during the Tang dynasty: Evidence from the Astana Cemetery. *Journal of Archaeological Science: Reports, 53*, 104298.

Cui, L., Chang, S. K., & Nannapaneni, R. (2021). Comparative studies on the effect of probiotic additions on the physicochemical and microbiological properties of yoghurt made from soymilk and cow's milk during refrigeration storage (R2). *Food Control, 119*, 107474.

Curl, S., Rivero-Mendoza, D., & Dahl, W. J. (2020). Plant-based milks: Hemp: FSHN20-53/ FS420, 10/2020. *Edis, 2020*(5).

da Silva, L. R., Velasco, J. I., & Fakhouri, F. M. (2023). Use of rice on the development of plant-based milk with antioxidant properties: From raw material to residue. *Lwt, 173*, 114271.

Dai, T., Shuai, X., Chen, J., Li, C., Wang, J., Liu, W., ... & Wang, R. (2022). Whole peanut milk prepared by an industry-scale microfluidization system: Physical stability, microstructure, and flavor properties. *Lwt, 171*, 114140.

Das, A., Raychaudhuri, U., & Chakraborty, R. (2012). Cereal based functional food of Indian subcontinent: A review. *Journal of Food Science and Technology, 49*, 665–672.

De, B., Shrivastav, A., Das, T., & Goswami, T. K. (2022). Physicochemical and nutritional assessment of soy milk and soymilk products and comparative evaluation of their effects on blood gluco-lipid profile. *Applied Food Research, 2*(2), 100146.

De Leon, S. Y., Delores, M. I., Barrett, D. M., Somogyi, L. P., & Ramaswamy. (2005). Coconut. In H. S. (Ed.), *Processing fruits: Science and technology*.

Deswal, A., Deora, N. S., & Mishra, H. N. (2014). Optimization of enzymatic production process of oat milk using response surface methodology. *Food and Bioprocess Technology, 7*, 610–618.

Devnani, B., Ong, L., Kentish, S., & Gras, S. (2020). Heat induced denaturation, aggregation and gelation of almond proteins in skim and full fat almond milk. *Food Chemistry, 325*, 126901.

Diarra, K., Zhang, G. N., & Chen, J. (2005). Peanut milk and peanut milk-based products production: A review. *Critical Reviews in Food Science and Nutrition, 45*(5), 405–423.

Dinu, M., Abbate, R., Gensini, G. F., Casini, A., & Sofi, F. (2017). Vegetarian, vegan diets and multiple health outcomes: A systematic review with meta-analysis of observational studies. *Critical Reviews in Food Science and Nutrition, 57*(17), 3640–3649.

Ecomil. Ecomil Quinoa drink sugar-free Bio 1 L. (2019). Retrieved from https://ecomil.com/en/productos/ecomil-quinoa-drink-sugar-free-bio-1-l/

Elmhurst 1925. Milked Oats. (2019). Retrieved from https://elmhurst1925.com/products/milked-oats

Fukui, K., Tachibana, N., & Wanezaki, S. (2002). Isoflavone free soy protein prepared by column chromatography reduces plasma cholesterol in rats. *Journal of Agriculture and Food Chemistry, 50*(20), 5717–5721.

Grasso, N., Alonso-Miravalles, L., & O'Mahony, J. A. (2020). Composition, physicochemical and sensorial properties of commercial plant-based yogurts. *Foods, 9*(3), 252.

Haas, R., Schnepps, A., Pichler, A., & Meixner, O. (2019). Cow milk versus plant-based milk substitutes: A comparison of product image and motivational structure of consumption. *Sustainability, 11*(18), 5046.

Harrington, L. K., & Mayberry, J. F. (2008). A re-appraisal of lactose intolerance. *International Journal of Clinical Practice, 62*(10), 1541–1546.

Heyman, M. B. (2006). Lactose intolerance in infants, children, and adolescents. *Pediatrics, 118*(3), 1279–1286.

Hsu, E. (2020). Plant-based diets and bone health: Sorting through the evidence. *Current Opinion in Endocrinology, Diabetes and Obesity, 27*(4), 248–252.

Ibrahim, S. A., & Gyawali, R. (2013). Lactose intolerance. In *Milk and dairy products in human nutrition: Production, composition and health* (pp. 246–260).

Karunasiri, A. N., Gunawardane, M., Senanayake, C. M., Jayathilaka, N., & Seneviratne, K. N. (2020). Antioxidant and nutritional properties of domestic and commercial coconut milk preparations. *International Journal of Food Science, 2020*(1), 3489605.

Kim, D. S., Iida, F., & Joo, N. (2022). Comparison of the nutritional composition of raw, boiled, roasted Macadamia (F. Muell. Macadamieae) and its application as an elder-friendly food. *International Journal of Gastronomy and Food Science, 30*, 100592.

Kittibunchakul, S., Yuthaworawit, N., Whanmek, K., Suttisansanee, U., & Santivarangkna, C. (2021). Health beneficial properties of a novel plant-based probiotic drink produced by fermentation of brown rice milk with GABAproducing Lactobacillus pentosus isolated from Thai pickled weed. *Journal of Functional Foods, 86*, 104710.

Leu, K. (1974). Analysis of volatile compounds produced in linoleic acid oxidation catalyzed by lipoxygenase from peas, soybeans and corn germs. *LWT Food Science and Technology, 7*, 98–100.

Liu, K. (2004). *Soybeans as functional foods and ingredients*. AOCS Press.

Maftei, N. M., Iancu, A. V., Elisei, A. M., Gurau, T. V., Ramos-Villarroel, A. Y., & Lisa, E. L. (2023). Functional characterization of fermented beverages based on soy milk and sea buckthorn powder. *Microorganisms, 11*(6), 1493.

Mäkinen, O. E., Wanhalinna, V., Zannini, E., & Arendt, E. K. (2016). Foods for special dietary needs: Non-dairy plant-based milk substitutes and fermented dairy-type products. *Critical Reviews in Food Science and Nutrition, 56*(3), 339–349.

Manzoor, M. F., Manzoor, A., Siddique, R., & Ahmad, N. (2017). Nutritional and sensory properties of cashew seed (Anacardiumoccidentale) milk. *Modern Concepts in Developmental Agronomy, 1*, 1–4.

Mattison, C. P., Aryana, K. J., Clermont, K., Prestenburg, E., Lloyd, S. W., Grimm, C. C., & Wasserman, R. L. (2020). Microbiological, physicochemical, and immunological analysis of a commercial cashew nut-based yogurt. *International Journal of Molecular Sciences, 21*(21), 8267.

McCarthy, K. S., Parker, M., Ameerally, A., Drake, S. L., & Drake, M. A. (2017). Drivers of choice for fluid milk versus plant-based alternatives: What are consumer perceptions of fluid milk?. *Journal of Dairy Science, 100*(8), 6125–6138.

McClements, D. J. (2020). Development of next-generation nutritionally fortified plant-based milk substitutes: Structural design principles. *Foods, 9*(4), 421.

Mobassaleh, M., Montgomery, R. K., Biller, J. A., & Grand, R. J. (1985). Development of carbohydrate absorption in the fetus and neonate. *Pediatrics, 75*(1), 160–166.

Moore, S. S., Costa, A., Pozza, M., Weaver, C. M., & De Marchi, M. (2024). Nutritional scores of milk and plantbased alternatives and their difference in contribution to human nutrition. *LWT, 191*, 115688.

Mygdalia, A., Sfetsas, T., Dimitropoulou, G., Zioupou, S., Mitsopoulos, T., Lithoxopoulos, P., ... & Katsantonis, D. (2023). Recipe for brown rice milk-based vegan ice cream. *Asian Food Science Journal, 22*, 33–39.

Nakamura, M., Kitamura, Y., & Kokawa, M. (2016). Development of a cheese-type food using rice milk. *Food Science and Technology Research, 22*(5), 605–609.

Namiki, M. (2007). Nutraceutical functions of sesame: A review. *Critical Reviews in Food Science and Nutrition, 47*(7), 651–673.

Nande, P., Tapadia, P., Jain, K., Lodhaya, F., & Vali, S. (2008). A study on soy milk as a substitute for animal milk. *Journal of Dairy Foods and Home Science, 27*(1), 1–10.

Ngamsuk, S., Hsu, J. L., Huang, T. C., & Suwannaporn, P. (2020). Ultrasonication of milky stage rice milk with bioactive peptides from rice bran: Its bioactivities and absorption. *Food and Bioprocess Technology, 13*, 462–474.

Niki, E., Yamamoto, Y., Takahashi, M., Komuro, E., & Miyama, Y. (1989). Inhibition of oxidation of biomembranes by tocopherol. *Annals of New York Academy of Sciences, 570*, 23–31.

Olías, R., Delgado-Andrade, C., Padial, M., Marín-Manzano, M. C., & Clemente, A. (2023). An updated review of soy-derived beverages: Nutrition, processing, and bioactivity. *Foods, 12*(14), 2665.

Omoni, A. O., & Aluko, R. E. (2005). Soybean foods and their benefits: Potential mechanisms of action. *Nutrition Reviews, 63*(8), 272–283.

Önning, G., Åkesson, B., Öste, R., & Lundquist, I. (1998). Effects of consumption of oat milk, soya milk, or cow's milk on plasma lipids and antioxidative capacity in healthy subjects. *Annals of Nutrition and Metabolism, 42*(4), 211–220.

Önning, G., Wallmark, A., Persson, M., Åkesson, B., Elmståhl, S., & Öste, R. (1999). Consumption of oat milk for 5 weeks lowers serum cholesterol and LDL cholesterol in free-living men with moderate hypercholesterolemia. *Annals of Nutrition and Metabolism, 43*(5), 301–309.

Patisaul, H. B., & Jefferson, W. (2010). The pros and cons of phytoestrogens. *Frontiers in Neuroendocrinology, 31*(4), 400–419.

Paul, A. A., Kumar, S., Kumar, V., & Sharma, R. (2019). Milk analog: Plant-based alternatives to conventional milk, production, potential and health concerns. *Critical Reviews in Food Science and Nutrition, 60*(18), 3005–3023.

Pointke, M., Ohlau, M., Risius, A., & Pawelzik, E. (2022). Plant-based only: Investigating consumers' sensory perception, motivation, and knowledge of different plant-based alternative products on the market. *Foods, 11*(15), 2339.

Prakash, V., & Narasinga Rao, M. S. (1986). Physicochemical properties of oilseed protein. *Critical Reviews in Biochemistry, 20*(3), 265–363.

Quasem, J. M., Mazahreh, A. S., & Abu-Alruz, K. (2009). Development of vegetable-based milk from decorticated sesame (SesamumIndicum). *American Journal of Applied Sciences, 6*, 888–896.

Rasane, P., Jha, A., Sabikhi, L., Kumar, A., & Unnikrishnan, V. S. (2015). Nutritional advantages of oats and opportunities for its processing as value-added foods—A review. *Journal of Food Science and Technology, 52*, 662–675.

Rasika, D. M., Vidanarachchi, J. K., Rocha, R. S., Balthazar, C. F., Cruz, A. G., Sant'Ana, A. S., & Ranadheera, C. S. (2021). Plant-based milk substitutes as emerging probiotic carriers. *Current Opinion in Food Science, 38*, 8–20.

Rasinperä, H., Kuokkanen, M., Kolho, K. L., Lindahl, H., Enattah, N. S., Savilahti, E., et al. (2005). Transcriptional downregulation of the lactase (LCT) gene during childhood. *Gut, 54*(11), 1660–1661.

Rizzo, G., Storz, M. A., & Calapai, G. (2023). The role of hemp (Cannabis sativa L.) as a functional food in vegetarian nutrition. *Foods, 12*(18), 3505.

Robayo-Torres, C. C., & Nichols, B. L. (2007). Molecular differentiation of congenital lactase deficiency from adult-type hypolactasia. *Nutrition Reviews, 65*(2), 95–98.

Romulo, A. (2022, February). Nutritional contents and processing of plant-based milk: A review. In *IOP Conference Series: Earth and Environmental Science* (Vol. 998, No. 1, p. 012054). IOP Publishing.

Sakthi, T. S., Meenakshi, V., Kanchana, S., & Vellaikumar, S. (2020). Study on standardisation and quality evaluation of peanut milk by different processing methods. *European Journal of Nutrition and Food Safety, 12*, 60–72.

Sathe, S. K., Wolf, W. J., Roux, K. H., Teuber, S. S., Venkatachalam, M., & Sze-Tao, K. W. C. (2002). Biochemical characterization of amandin, the major storage protein in almond (Prunusdulcis L.). *Journal of Agriculture and Food Chemistry, 50*(15), 4333–4341.

Segal, S., & Blair, A. (1961). Some observations on the metabolism of D-galactose in normal man. *The Journal of Clinical Investigation, 40*(11), 2016–2025.

Seow, C. C., & Gwee, C. N. (1997). Coconut milk: Chemistry and technology. *International Journal of Food Science and Technology, 32*, 189–201.

Sethi, S., Tyagi, S. K., & Anurag, R. K. (2016). Plant-based milk alternatives: An emerging segment of functional beverages. *Journal of Food Science and Technology, 53*, 3408–3423.

Settaluri, V. S., Kandala, C. V. K., Puppala, N., & Sundaram, J. (2012). Peanuts and their nutritional aspects—A review. *Food and Nutrition Sciences, 3*, 1644–1650.

Sevillano Pires, V., Zuklic, J., Hryshko, J., Hansen, P., Boyer, M., Wan, J., ... & Redan, B. W. (2022). Market basket survey of the micronutrients vitamin A, vitamin D, calcium, and potassium in eight types of commercial plant-based milk alternatives from United States markets. *ACS Food Science & Technology, 3*(1), 100–112.

Shori, A. B., Aljohani, G. S., Al-zahrani, A. J., Al-sulbi, O. S., & Baba, A. S. (2022). Viability of probiotics and antioxidant activity of cashew milk-based yogurt fermented with selected strains of probiotic Lactobacillus spp. *Lwt, 153*, 112482.

Suchy, F. J., Brannon, P. M., Carpenter, T. O., Fernandez, J. R., Gilsanz, V., Gould, J. B., & Mennella, J. (2010). National Institutes of Health consensus development conference: Lactose intolerance and health. *Annals of Internal Medicine, 152*, 792–796.

Tangyu, M., Muller, J., Bolten, C. J., & Wittmann, C. (2019). Fermentation of plant-based milk alternatives for improved flavour and nutritional value. *Applied Microbiology and Biotechnology, 103*, 9263–9275.

Tomova, A., Bukovsky, I., Rembert, E., Yonas, W., Alwarith, J., Barnard, N. D., & Kahleova, H. (2019). The effects of vegetarian and vegan diets on gut microbiota. *Frontiers in Nutrition, 6*, 47.

Topcuoglu, E., & Yilmaz-Ersan, L. (2020). Effect of fortification with almond milk on quality characteristics of probiotic yoghurt. *Journal of Food Processing and Preservation, 44*(12), e14943.

Tulashie, S. K., Amenakpor, J., Atisey, S., Odai, R., & Akpari, E. E. A. (2022). Production of coconut milk: A sustainable alternative plant-based milk. *Case Studies in Chemical and Environmental Engineering, 6*, 100206.

United States Department of Agriculture, USDA. (2019). *USDA nutrient database for standard reference, release*. Nutrient Data Laboratory Home Page. Retrieved from http://ndb.nal.usda.gov Accessed 17 Jan 2019.

Vaikma, H., Kaleda, A., Rosend, J., & Rosenvald, S. (2021). Market mapping of plant-based milk alternatives by using sensory (RATA) and GC analysis. *Future Foods, 4*, 100049.

Vanga, S. K., & Raghavan, V. (2018). How well do plant-based alternatives fare nutritionally compared to cow's milk? *Journal of Food Science and Technology, 55*(1), 10–20.

Wang, R., Zhou, X., & Chen, Z. (2008). High-pressure inactivation of lipoxygenase in soy milk and crude soybean extract. *Food Chemistry, 106*(2), 603–611.

Welch, R. W. (1995). Oats in human nutrition and health. In *The oat crop: Production and utilization* (pp. 433–479). Springer.

Wien, M., Oda, K., & Sabaté, J. (2014). A randomized controlled trial to evaluate the effect of incorporating peanuts into an American Diabetes Association meal plan on the nutrient profile of the total diet and cardiometabolic parameters of adults with type 2 diabetes. *Nutrition Journal, 13*, 1–9.

Yamamoto, M., Handa, N., Nakamura, A., Takahashi, H., & Kuda, T. (2022). In vitro antioxidant, anti-glycation, and bile acid-lowering capacity of peanut milk fermented with Lactiplantibacillus plantarum Kinko-SU4. *Current Research in Food Science, 5*, 992–997.

Yilmaz-Ersan, L., & Topcuoglu, E. (2022). Evaluation of instrumental and sensory measurements using multivariate analysis in probiotic yogurt enriched with almond milk. *Journal of Food Science and Technology*, 1–11.

Yu, Y., Li, X., Zhang, J., Li, X., Wang, J., & Sun, B. (2023). Oat milk analogue versus traditional milk: Comprehensive evaluation of scientific evidence for processing techniques and health effects. *Food Chemistry: X*, 100859.

Yuan, H., Luo, Z., Ban, Z., Reiter, R. J., Ma, Q., Liang, Z., & Li, L. (2022). Bioactive peptides of plant origin: Distribution, functionality, and evidence of benefits in food and health. *Food & Function, 13*(6), 3133–3158.

Zahra, A. K., Varidi, M., Varidi, M. J., & Pourazarang, H. (2014). Influence of processing conditions on the physicochemical and sensory properties of sesame milk: A novel nutritional beverage. *LWT Food Science and Technology, 57*(1), 299–305.

Zahrani, A. J. A., & Shori, A. B. (2023). Viability of probiotics and antioxidant activity of soy and almond milk fermented with selected strains of probiotic Lactobacillus spp. *LWT, 176*, 114531.

Zor, M., Bulut, M., Karagöz, S. G., Çetintaş, Y., & Alwazeer, D. (2024). Use of Hydrogen-Rich water in rice milk preparation improves the nutritional and sensory properties of product. *Food Chemistry, 437*, 137821.

Zhou, S., Jia, Q., Cui, L., Dai, Y., Li, R., Tang, J., & Lu, J. (2023). Physical–chemical and sensory quality of oat milk produced using different cultivars. *Foods, 12*(6), 1165.

Chapter 8
Health and Regulatory Aspects of Lactose Hydrolyzed Dairy Products

Richa Singh, Rajan Sharma, Sumit Arora, and Anusha Kishore

8.1 Introduction

Lactose, a disaccharide is only found in milk and milk products. Lactose intolerance is caused by the declined activity or absence of β-galactosidase enzyme (also known as lactase) in gastrointestinal tract. As per recent report, 65% of world population and 95% of Asian population is lactose intolerance (Li et al., 2023). In a comprehensive review and meta-analysis conducted by Storhaug et al. (2017), which incorporated data from both MEDLINE and Embase. It encompassed 62,910 participants hailing from 89 countries and established that lactose malabsorption has a global prevalence of 68%. This also reaffirmed the extensive occurrence of lactose malabsorption on a worldwide scale. The traditional treatment of lactose intolerance is avoidance of dietary lactose in diet of an individual. However, evolving landscape of dietary solutions emphasizes the nutraceutical and functional aspects of foods. This includes creating dairy products with little or no lactose. Moreover, the development of such dairy products is an effective strategy to address lactose intolerance as it enables such individuals to consume crucial nutrients included in milk. Lactose free milk, flavored milk, cheese, ice-cream, yoghurt, powders etc. are some products available in the market as an option for lactose intolerant population.

R. Singh (✉) · R. Sharma · S. Arora · A. Kishore
Dairy Chemistry Division, ICAR-National Dairy Research Institute, Karnal, Haryana, India

P. S. Rao et al. (eds.), *Lactose Hydrolysis in Dairy Products*,
https://doi.org/10.1007/978-3-031-78207-7_8

8.2 Nutritional Difference Between Regular Milk Products and Lactose Free Milk Products

Lactose-free dairy products allow lactose intolerant individuals to enjoy the taste and nutrition of milk without the uncomfortable intestinal symptoms. The technology behind eliminating lactose primarily involves either breakdown of lactose through microbial β-galactosidase or physical removal of lactose (e.g., via processes like ultrafiltration or chromatography), and sometimes a combination of both methods is employed. In the process where β-galactosidase is used for lactose removal, the lactose content may be very low (<0.01 g/100 mL), but the glucose and galactose content of the prepared products will be approximately 2.5 g/100 mL. Thus, this lactose hydrolyzed products are not suitable for people suffering from mL. Thus, people suffering from galactosemia may consume lactose free milk prepared using membrane filtration because of the absence of these monomers. Glucose and galactose produced on lactose hydrolysis are sweeter than lactose. Hence, manufacture of lactose free milk products from lactose free milk allows for a reduction in the amount of added sugar by as much as 10–15 grams per kilogram of final product. This reduction in added sugar content contributes to a decrease in total calorie intake. Again, it is presumed that lactose hydrolyzed milk is not suitable for diabetic persons due to a higher level of monosaccharides glucose and galactose that may increase blood sugar level. However, findings also suggest that on ingestion of predigested lactose, glucose and galactose are absorbed similar to intact lactose in the small intestine. Ercan et al. (1993) confirmed this absorption pattern among diabetes patients and reported non-significant differences in glycemic response on consumption of lactose containing product or lactose hydrolyzed product containing glucose and galactose. Furthermore, da-Costa-Pinto and Collares (1997) also found no disparities in gastric emptying when comparing the consumption of lactose to that of glucose and galactose using an animal model (rats). Similarly, Gutzwiller (2000) also reported no variance in glycemic responses between regular milk and lactose-free milk.

8.3 Health Aspects of Lactose-Hydrolyzed Milk and Milk Products

The complete avoidance of lactose consumption by lactose intolerant population with the use of plant-based products seems straightforward, but it will not be as nutritious as lactose free milk products. Nutritional deficiencies were observed when milk-based diets were avoided (Wahlqvist, 2015). Eliminating all dairy products from one's diet can potentially lead to nutrient deficiencies since these foods serve as a primary source of essential nutrients, especially calcium, phosphorus, riboflavin, vitamin B12, and vitamin A. Several studies have suggested that the omission of dairy from diet of individuals is linked to nutritional deficiencies and decreased bone mineral density (Laaksonen et al., 2009; Heaney, 2013). Conversely, sufficient dairy consumption is indicative of a nutritionally rich diet (Nicklas et al., 2009; Heaney, 2013) and is associated with a reduced risk of osteoporosis, diabetes, colorectal cancer, hypertension and weight gain (Imamura et al., 2018; Mozaffarian & Wu, 2018). Thus, limiting intake of lactose by consuming lactose-reduced or lactose free dairy products is a wise decision, as nutritional effects of lactose-free milk products is similar to regular dairy (Dekker et al., 2019). The usage of popular plant-based drinks as a dietary management for lactose intolerance has been reported to cause calcium deficiency. According to the findings from EPIC-Oxford cohort study conducted by Key et al. (2007) involving 34,696 individuals from UK, it was observed that vegans faced a 30% greater risk of bone fractures compared to omnivores and vegetarians (those who include milk, yogurt, cheese, and eggs in their diet). This increased risk was primarily linked to lower calcium intake (Tong et al., 2020). The calcium content of lactose free milk and some popular plant-based drinks is presented in Table 8.1. If plant-based diets are opted for lactose-free products, they must be fortified with at least 20% more than the recommended intake for calcium due to 70-75% bioavailability from plant-based diets.

Calcium is an essential component for mineralization of teeth and bones as well as for controlling intracellular processes. The recommended daily allowance (RDAs) for calcium is 800 mg (Indian Council of Medical Research, 2020) and the requirement is almost double for lactating and pregnant women. A lack of calcium in diet due to avoidance of milk and dairy products can lead to osteoporosis and fractures later in life (Wilt et al., 2010).

Absorption of calcium from low-lactose or lactose free milk and from regular milk may be different and reports are available in literature pointing towards enhanced absorption of minerals like calcium in presence of lactose. Brommage et al. (1993) observed a rise in calcium absorption from 17 to 30 % with an increase in dietary lactose content from 0 % to 15% in a study conducted on rats. However, clinical trials related to the effect of lactose on absorption of calcium have revealed conflicting findings. In some studies, calcium absorption was shown to increase with higher lactose content in infant diets (Ziegler & Fomon, 1983), while in others the influence of lactose on calcium absorption appeared to be minimal (Wirth Jr et al., 1990). It has been proposed that the conflicting findings can be attributed to variations in the choice of control sugars used in experiments. When glucose or galactose were used as control sugars, they yielded similar ($p>0.05$) calcium

Table 8.1 Fat, Protein, carbohydrate and calcium content of lactose hydrolyzed milk and plant-based drinks (Suri et al., 2019)

Sr. no.	Type of milk	Fat (%)	Protein (%)	Carbohydrate (%)	Calcium(ppm)
1.	Lactose hydrolysed milk	1.5–4.3	3.20–3.37	4.8	112–170
2.	Soy milk	1.5	3.0	1.5	3.90
3.	Rice milk	0.79–2.5	15.5	45–57.3	1.50–1.60
4.	Coconut milk	32.2–40	2.8–4.4	5.5–8.3	9.40
5.	Oat milk	1.30–1.52	0.7–1.0	12	–

absorption rates as lactose. However, when non-absorbable sugars used as control, lactose did demonstrate a positive effect ($P < 0.05$) on calcium absorption in humans. Unhydrolyzed lactose appears to function as a prebiotic, supporting the growth of beneficial gut bacteria recognized for their role in enhancing calcium absorption. However, both hydrolyzed and unhydrolyzed forms of lactose seem to play a positive role in augmenting calcium absorption in humans (Kwak et al., 2012).

Milk proteins contain all essential amino acid in appreciable amount and hence the quality of milk proteins is better than plant proteins. The protein efficiency ratio, biological value and net protein utilization of milk protein is 3.1, 91 and 82 which are better than soy protein for which these values are 2.1, 74 and 64, respectively. Milk was found to be outstanding when compared with plant-based drinks (soy, almond, cashew, coconut and oat) and exhibited higher calculated DIAASs (Digestible Indispensable Amino acid score) (Walther et al., 2022).

8.4 Market for Lactose Hydrolyzed Milk and Milk Products

According to the National Institute of Health (NIH), USA, most individuals with lactose malabsorption can comfortably handle up to 12 grams of lactose in a single serving (Suchy et al., 2010). Furthermore, various factors such as the frequency and quantity of lactose consumption, intestinal flora, gastrointestinal motility, and sensitivity to lactose metabolites all contribute to the clinical symptoms. Lactose-free dairy products have emerged as the dairy industry's most rapidly growing segment. The global compound annual growth rate for lactose-free dairy products between 2017 and 2022 stood at approximately 7%, constituting a significant portion, amounting to 80%, of all lactose-free product sales (Dekker et al., 2019). Over the past 5 years, the variety of lactose-free dairy offerings has continued to expand. Among these, lactose-free milk and yogurt have emerged as the most prevalent options, while lactose-free butter, cheese, and flavored milk have also gained popularity. In different countries different lactose free dairy products are occupying their space on shelf. Notably, Valio, Finland's largest food and dairy company achieved a significant milestone by becoming the world's first producer of lactose-free dairy goods, introducing lactose-free milk in 2001. Świąder et al. (2020) conducted an analysis of lactose-free dairy products market in Poland and categorized a total of 75 lactose-free dairy items. These encompassed a wide range of products, including milk, yogurt, cream, quark, cheese, and infant formula. Arla Foods in United Kingdom is also well-recognized for Lactofree brand. Fairlife, an American company, has also launched a range of lactose-free ultrafiltered milk products, such as shakes and recovery drinks. These products are notably rich in calcium and proteins, securing a strong presence in this expanding market segment. In India, Amul has launched lactose free milk in the year 2015. However, at present few local firms have also joined this segment like Provilac Dairy Farm, Pune; Milky Mist Dairy Foods Pvt Ltd. Tamil Nadu but the availability of these products is limited.

Table 8.2 Low-lactose and lactose free products available on shelf in different countries

Sr. no.	Country	Products	Lactose content mentioned on the label[a]
1.	Finland	Valio Eila® Pro lactose free milk powder	<0.1%
2.		Valio Eila® NUTRI F+ lactose free milk protein powder	<0.05%
3.		Valio play® lactose-free yoghurt	<0.01
4.		Valio play® lactose-free drink	0
5.		Valio Play® cool® fruit bomb lactose-free	0
6.	United Kingdom	Arla lactofree milk, yoghurt, cheese, cheese spread, cream	On package mentioned lactose free but lactose content is not mentioned.
7.	France	Lactalis Lactose free skim milk powder	<0.1%
8.		Lactalis Low-lactose skimmed milk powder	<1.5%
9.		Lactalis Native low-lactose whey protein isolates Pronativ®	0.2%
10.		Danone Activia lactose free yoghurt and yoghurt drink	On package mentioned lactose free but lactose content is not mentioned.
11.		Yoplait® lactose free yoghurt	On package mentioned lactose free but lactose content is not mentioned.
12.	United States of America	Fairlife® ultrafiltered milk, flavoured milk	No lactose is mentioned on label
13.		Green valley lactose free yoghurt, cheese, kefir, cream	On package mentioned lactose free but lactose content is not mentioned.
14.	India	Amul lactose free milk	<0.01%
15.		Provilac lactose free milk	On package mentioned lactose free but lactose content is not mentioned.
16.		MilkyMist® lactose free milk	Nil mentioned on label

[a]The information is compiled from open source

It can be safely anticipated (Table 8.2) that lactose content of low-lactose and lactose-free products available in market are as low as 0.01% to 1.5%. Regulations for lactose free dairy products are still in a nascent stage (Table 8.3) and except for few standard products; lactose content is not mentioned on the label. Therefore, consumers are not able to make informed decisions. Since, plant-based products are marketed with claim of 100 % lactose free, they are gaining popularity among lactose intolerant population. Therefore, it becomes important to have regulatory standards for lactose free dairy products so that lactose intolerant consumers feel safe in consuming such products. As a part of regulatory standards, only few countries have prescribed standards for low-lactose and lactose free milk and milk products (Table 3) but still the labelling part is not strictly regulated.

Table 8.3 Legal standards applicable in different countries for low lactose and lactose free products.

Sr. no.	Country	Lactose content	Regulatory agency
1.	Denmark	<10 mg/100 g as lactose free <1 g/100 g as low lactose	Danish Agriculture and Food Council
2.	Finland	<10 mg/100 g as lactose free <1 g/100 g as low lactose	Finnish Food Authority
3.	Sweden	<10 mg/100 g as lactose free <1 g/100 g as low lactose	Swedish Food Agency
4.	Norway	<10 mg/100 g as lactose free <1 g/100 g as low lactose	Norwegian Ministry of Agriculture and Food
5.	Germany	<100 mg/100 g as lactose free	Federal Office of Consumer Protection and Food Safety
6.	Hungary	<100 mg/100 g as lactose free	Hungarian Food Act
7.	Ireland	No lactose present as lactose <1 g/100 g as low lactose	Food Safety Authority of Ireland
8.	India	<1000 mg/100 mg as low lactose	Food Safety and Standards Authority of India
9.	India	< 10 mg/100 g as lactose free	
10.	China	<500 mg/100 g as lactose free	National Health Commission
11.	Italy	<100 mg/100 g as lactose free	Italian Health Ministry
12.	Switzerland	<100 mg/L as lactose free	Swiss Federal Food Safety and Veterinary Office
13.	Kenya	<1000 mg/100 g as low lactose	Kenya Dairy Board

European Food Safety Authority (2010), Food Safety and Standards regulation (2019) and Kenya Bureau of Standards, DKS 2934: 2021 (2020)

8.5 Regulatory Aspect of Low Lactose and Lactose Free Products in European Union and India

European countries are leading in the market for lactose free products. In the European Community legislation, the term "lactose-free" has only been defined for infant and follow-on formula as < 2.5 mg/100 kJ or ≤10 mg/100 kcal (EU Regulation 609/2013). In European Union, plant-based drinks are also gaining popularity among the lactose intolerant population. To differentiate plant-based products from lactose free dairy products, the lactose free dairy products are marketed under the denomination 'milk-based lactose free drinks. Some EU member states have set threshold levels at national level for the use of terms "lactose free", "very low lactose" and "low lactose" for foodstuffs other than those intended for infants (as mentioned in Table 8.3).

In India, Food Safety and Standards Authority of India (FSSAI) is the regulatory authority for milk and milk products including lactose free products. As per Food Safety and Standards Regulation (FSSR), 2019; low lactose or lactose free milk means the product prepared from any type of milk (species specific milk, mixed milk, standardized milk, full cream milk, toned milk, double toned milk or skim milk) in which, lactose content has been reduced significantly through hydrolysis by enzymatic or any other appropriate process. It will be called "low lactose milk" if it has lactose content less than 1 %; and "lactose free milk" if it has lactose content less than 0.1% lactose. The fat and SNF content of milk used for preparation shall conform to the respective composition defined by FSSR. The product may be subjected to pasteurization, boiling, sterilisation or ultra-high temperature.

8.6 Labelling Requirements of Lactose Free Products

In India, as per FSSR the lactase enzyme is required to be mentioned under the list of ingredients for lactose hydrolyzed products and following declarations are also required on the label of package:

1. Low lactose or lactose free......................milk, wherein the blank will be filled by the name of the type milk from which it is prepared.
2. The heat treatment to which product has been subjected to.
3. If the milk from any milch animal, mixed milk or skimmed milk is offered for sale to the consumer without any heat treatment, the name of the milk shall be declared on the label of pre-packaged milk;

Food Safety and Standards (Packaging and labelling) Regulations (2011), the container of infant milk substitute for lactose intolerant infants should be labeled in capital letters as "LACTOSE-FREE and statement "TO BE TAKEN UNDER MEDICAL ADVICE". It should bear the following statements:

"Lactose free infant milk substitute should only be used in case of diarrhea due to lactose intolerance. The lactose free infant milk substitute should be withdrawn if there is no improvement in symptoms of intolerance".

According to the United States Food and Drug Administration (USFDA), lactose-free products must not contain any lactose at all, and reduced lactose products must be labelled as reduced lactose and not lactose-free. However, it is not necessary to mention lactose content on the label and hence the products available in US market are not mentioning lactose content on the label.

In **European Union (EU)**, the labelling of lactose-free products is regulated under article 36 of **EU** Regulation 1169/2011 on Food Information to Consumers. On all the products, lactose content is required to be mentioned on the label. In lactose hydrolyzed dairy products, lactase is required to be mentioned in the list of ingredients and since these products will have a large quantity of galactose a warning has to be included "not suitable for people with galactosaemia". When the statement 'lactose free' is used for infant formula, it shall be accompanied with the statement 'not suitable for infants with galactosaemia', which shall be indicated with the same font size and prominence as the statement 'lactose free' and in close proximity to it."

In European Union, the panel on Dietetic Products, Nutrition and Allergies has provided a scientific opinion regarding health claims related to milk and milk products prepared by lactase enzyme (EFSA, 2010). Since a lot of these products are available in European market, member countries have been requested to use following two health claims on such products.

1. It helps lactose digestion by reducing fermentation and gas production.
2. Breaks down lactose. Useful for people whose own lactase enzyme production is insufficient for breaking down lactose.

The broad claimed health effect for such type of products is related to 'digestion'. This claim is intended for individuals who have an insufficient natural lactase enzyme production to effectively break down lactose. Enhancing the breakdown of lactose can be advantageous for the well-being of those with symptomatic lactose maldigestion. Scientific evidence has established a clear cause-and-effect relationship between the consumption of lactase enzyme and breakdown of lactose in individuals with symptomatic lactose maldigestion. Therefore, the phrase "lactase enzyme contributes to breaking down lactose" has been approved for use on lactose free products.

8.7 Confirmation of Lactose Content Claims on Lactose Free or Low Lactose Milk Products

Various methods have been employed to confirm labeled lactose concentrations in dairy products, with challenges in distinguishing carbohydrates and enzymatically reduced lactose. Enzymatic approaches hydrolyze lactose to glucose and galactose, measuring the difference before and after, often utilizing β-galactose dehydrogenase (AOAC, 2007; ISO and IDF, 2002a, b). These methods offer a quantification limit as low as 0.01 g/100 g, most accurate at higher lactose concentrations. Mangan et al. (2019) developed an innovative method LOLAC (low lactose) for detection and estimation of lactose in lactose free milk and milk products samples. The process begins with an efficient pre-treatment step for glucose removal, followed by a sequential enzymatic assay that measures the remaining levels of glucose and lactose in a single cuvette. The method's limit of detection (LOD) and limit of quantification (LOQ) were determined to be 0.13 mg/100 g and 0.44 mg/100 g, respectively, which is very much lower to the enzymatic assay kits commercially available.

High-Performance Liquid Chromatography (HPLC) directly detects carbohydrates, with refractive index (RI) detector having a sensitivity limit of 0.05 g/100 mL. Light-scattering detectors and electrochemical detectors enhance sensitivity. Reverse-phase and cationic exchange resin columns are commonly used.

High-performance anion-exchange chromatography (HPAEC) with pulsed amperometric detection (PAD) offers high sensitivity for nonderivatized carbohydrates, detecting lactose below 0.01 g/100 g. Capillary electrophoresis with electrochemical detection (CE-ED) can detect lactose at 0.01 mg/100 mL.

ISO 22662, IDF 198 (2007) is the reference method for lactose determination in raw milk, heat-treated milks, dried milk, and cream. However, this method which is a HPLC based method is not suitable for fermented milks or milks with added oligosaccharides. Further, its efficacy in estimation of residual lactose in low lactose products has not been established.

8.8 Conclusion

Low-lactose and lactose free milk and milk products are the best solution for lactose intolerant people. These products allow lactose intolerant people to enjoy the nutritional virtues of milk. The market for such products is increasing with consumer awareness and a lot of such products are gaining entry in the market. Regulatory agencies have recognized the presence of these products and have issued standards and labeling criteria for lactose free milk and milk products to enable the consumer to make informed choices. However, labelling on some available products in the market is confusing as it mentions lactose free but does not mention lactose content. Wordings related to health claims have also not been addressed by regulatory agencies except for European Union.

References

AOAC. (2007). Official methods of analysis of AOAC international.

Brommage, R., Binacua, C., Antille, S., & Carrié, A. L. (1993). Intestinal calcium absorption in rats is stimulated by dietary lactulose and other resistant sugars. *The Journal of nutrition, 123*(12), 2186–2194.

da-Costa-Pinto, E. A. L., & Collares, E. F. (1997). Chronic lactose intake modifies the gastric emptying of monosaccharides but not of disaccharides in weanling rats. *Brazilian journal of medical and biological research, 30*, 723–726.

Dekker, P. J., Koenders, D., & Bruins, M. J. (2019). Lactose-free dairy products: Market developments, production, nutrition and health benefits. *Nutrients, 11*(3), 551.

DKS 2934:2021. (2020). *Lactose free milk-specification*. Kenya Bureau of Standards.

EFSA Panel on Dietetic Products, Nutrition and Allergies (NDA). (2010). Scientific opinion on lactose thresholds in lactose intolerance and galactosaemia. *EFSA Journal, 8*(9), 1777.

Ercan, N., Nuttall, F. Q., Gannon, M. C., Redmon, J. B., & Sheridan, K. J. (1993). Effects of glucose, galactose, and lactose ingestion on the plasma glucose and insulin response in persons with non-insulin-dependent diabetes mellitus. *Metabolism, 42*(12), 1560–1567.

Food Safety and Standards Regulations (FSSR). (2019). *Food safety and standards (Food products standards and food additives) regulations, 2011*. Version-XXVI (20.12.2022). Food Safety Standards Authority of India.

Gutzwiller, A. (2000). Glucose and galactose absorption after ingestion of milk containing hydrolysed lactose in calves with diarrhoea. *Journal of Veterinary Medicine Series A, 47*(8), 495–500.
Heaney, R. P. (2013). Dairy intake, dietary adequacy, and lactose intolerance. *Advances in nutrition, 4*(2), 151–156.
ICMR. (2020). https://drklbcollege.ac.in/wp-content/uploads/2020/03/DOC-20220614-WA0002_-1.pdf. Indian Council of Medical Research.
Imamura, F., Fretts, A., Marklund, M., Ardisson Korat, A. V., Yang, W. S., Lankinen, M., Qureshi, W., Helmer, C., Chen, T. A., Wong, K., & Bassett, J. K. (2018). Fatty acid biomarkers of dairy fat consumption and incidence of type 2 diabetes: A pooled analysis of prospective cohort studies. *PLoS Medicine, 15*(10), e1002670.
ISO and IDF (International Organization for Standardization and International Dairy Federation), (2002a). Dried milk, dried ice-mixes and processed cheese - Determination of lactose content - Part 1: Enzymatic method utilizing the glucose moiety of the lactose.
ISO and IDF (International Organization for Standardization and International Dairy Federation), (2002b). Dried milk, dried ice-mixes and processed cheese - Determination of lactose content - Part 2: Enzymatic method utilizing the galactose moiety of the lactose.
ISO and IDF (International Organization for Standardization and International Dairy Federation), (2007). International standard: Milk and milk products – Determination of lactose content by high performance liquid chromatography (Reference method).
Key, T. J., Appleby, P. N., Spencer, E. A., Roddam, A. W., Neale, R. E., & Allen, N. E. (2007). Calcium, diet and fracture risk: a prospective study of 1898 incident fractures among 34 696 British women and men. *Public health nutrition, 10*(11), 1314–1320.
Kwak, H. S., Lee, W. J., & Lee, M. R. (2012). Revisiting lactose as an enhancer of calcium absorption. *International Dairy Journal, 22*(2), 147–151.
Laaksonen, M. M., Impivaara, O., Sievänen, H., Viikari, J. S., Lehtimäki, T. J., Lamberg-Allardt, C. J., Kärkkäinen, M. U., Välimäki, M., Heikkinen, J., Kröger, L. M., & Kröger, H. P. (2009). Associations of genetic lactase non-persistence and sex with bone loss in young adulthood. *Bone, 44*(5), 1003–1009.
Li, A., Zheng, J., Han, X., Yang, S., Cheng, S., Zhao, J., Zhou, W., & Lu, Y. (2023). Advances in low-lactose/lactose-free dairy products and their production. *Foods, 12*(13), 2553.
Mangan, D., McCleary, B. V., Culleton, H., Cornaggia, C., Ivory, R., McKie, V. A., Delaney, E., & Kargelis, T. (2019). A novel enzymatic method for the measurement of lactose in lactose-free products. *Journal of the Science of Food and Agriculture, 99*(2), 947–956.
Mozaffarian, D., & Wu, J. H. (2018). Flavonoids, dairy foods, and cardiovascular and metabolic health: A review of emerging biologic pathways. *Circulation research, 122*(2), 369–384.
Nicklas, T. A., Qu, H., Hughes, S. O., Wagner, S. E., Foushee, H. R., & Shewchuk, R. M. (2009). Prevalence of self-reported lactose intolerance in a multiethnic sample of adults. *Nutrition Today, 44*(5), 222–227.
Storhaug, C. L., Fosse, S. K., & Fadnes, L. T. (2017). Country, regional, and global estimates for lactose malabsorption in adults: a systematic review and meta-analysis. *The Lancet Gastroenterology & Hepatology, 2*(10), 738–746.
Suchy, F. J., Brannon, P. M., Carpenter, T. O., Fernandez, J. R., Gilsanz, V., Gould, J. B., Hall, K., Hui, S. L., Lupton, J., Mennella, J., Miller, N. J., & Wolf, M. A. (2010). National institutes of health consensus development conference: Lactose intolerance and health. *Annals of internal medicine, 152*(12), 792–796.
Suri, S., Kumar, V., Prasad, R., Tanwar, B., Goyal, A., Kaur, S., Gat, Y., Kumar, A., Kaur, J., & Singh, D. (2019). Considerations for development of lactose-free food. *Journal of Nutrition & Intermediary Metabolism, 15*, 27–34.
Świąder, K., Kulawiak, M., & Chen, Y. P. (2020). Types of lactose-free products and their availability on the Polish market®. *PostępyTechnikiPrzetwórstwaSpożywczego*.
Tong, T. Y., Appleby, P. N., Armstrong, M. E., Fensom, G. K., Knuppel, A., Papier, K., Perez-Cornago, A., Travis, R. C., & Key, T. J. (2020). Vegetarian and vegan diets and risks of total and site-specific fractures: Results from the prospective EPIC-Oxford study. *BMC medicine, 18*(1), 1–15.

Wahlqvist, M. L. (2015). Lactose nutrition in lactase nonpersisters. *Asia Pacific journal of clinical nutrition, 24*(Supplement 1), S21–S25.

Walther, B., Guggisberg, D., Badertscher, R., Egger, L., Portmann, R., Dubois, S., Haldimann, M., Kopf-Bolanz, K., Rhyn, P., Zoller, O., Veraguth, R., & Rezzi, S. (2022). Comparison of nutritional composition between plant-based drinks and cow's milk. *Frontiers in nutrition, 9*, 2645.

Wilt, T. J., Shaukat, A., Shamliyan, T., Taylor, B. C., MacDonald, R., Tacklind, J., Rutks, I., Schwarzenberg, S. J., Kane, R. L., & Levitt, M. (2010). Lactose intolerance and health. *Evidence report/technology assessment, 192*, 1–410.

Wirth, F. H., Jr., Numerof, B., Pleban, P., & Neylan, M. J. (1990). Effect of lactose on mineral absorption in preterm infants. *The Journal of pediatrics, 117*(2), 283–287.

Ziegler, E. E., & Fomon, S. J. (1983). Lactose enhances mineral absorption in infancy. *Journal of pediatric gastroenterology and nutrition, 2*(2), 288–294.

Chapter 9
Conversion of Milk Lactose into Other Derivatives

B. Ashritha, N. Veena, Avinash C. Gautam, and Nitika Goel

9.1 Introduction

As cheese production keeps growing and the practice of disposing of whey permeate solids in wastewater is increasingly seen as wasteful. Dairy manufacturers have seven options to make use of the permeate streams from milk and whey (both sweet and acid): drying, fermenting into different products (such as ethanol, lactic acid, and bacteriocins), producing lactose, standardizing protein in milk powders (using either milk permeate or lactose derived from whey permeate), hydrolyzing lactose, recovering minor components, and synthesizing derivatives from a lactose-rich substrate. Due to its current insufficient utilization, there is a substantial emphasis on research and developments of whey permeate and components produced from lactose (Ganzle et al., 2008).

It is possible to biotransform lactose into various monosaccharides or their derivatives by hydrolysis, transglycosylation, transgalactosylation, isomerization, and epimerization (Chen et al., 2018). The majority of lactose derivatives, such as galacto-oligosaccharides, epilactose, lactulose, lactosucrose, and D-tagatose, are useful carbohydrates and occur in nature in small quantities. β-galactosidase (EC 3.2.1.23) facilitates the breakdown of lactose into D-galactose and D-glucose by hydrolysis. Subsequently, L-arabinose-ketol-isomerase (EC 5.3.1.4) derived from

B. Ashritha · A. C. Gautam
Dairy Technology Section, Southern Regional Station, ICAR-National Dairy Research Institute, Bengaluru, Karnataka, India

N. Veena (✉)
Department of Food Safety and Quality, DSLD College of Horticultural Engineering and Food Technology, UHSB, Haveri, Karnataka, India
e-mail: veena.n@uhsbagalkot.edu.in

N. Goel
Department of Dairy Technology, College of Dairy and Food Science Technology, GADVASU, Ludhiana, Punjab, India

P. S. Rao et al. (eds.), *Lactose Hydrolysis in Dairy Products*,
https://doi.org/10.1007/978-3-031-78207-7_9

Lactobacillus gayonii transforms D-galactose into D-tagatose. β-Galactosidase catalyzes the transfer of the galactosyl group from lactose to produce galacto-oligosaccharides (GOS) (Lu et al., 2020). The sugar alcohols formed from lactose, such as lactitol, sorbitol, galactitol, and lactulitol, are conventionally produced by breaking down lactose into its components through hydrolysis and then subjecting them to hydrogenation. Lactitol, a less cariogenic sweetener, is the main product obtained from the hydrogenation of lactose. A tiny quantity of lactulitol is also present simultaneously. Furthermore, lactitol has the potential to undergo further hydrolysis and hydrogenation processes, resulting in the formation of sorbitol and galactitol (Martínez-Monteagudo et al., 2019). Lactulose (4-O-β-D-galactopyranosyl-D-fructofuranose) is a synthetic and non-digestible isomerized product derived from lactose. It consists of two disaccharide molecules, fructose and galactose (Aït Aissa & Aïder, 2014). Identical to undigested lactose, these compounds improve the absorption of calcium and magnesium in the intestines. The increasing economic interest in health-promoting functional food has focused on the possible health benefits of lactose derivatives. The physiological effects of constipation and hepatic encephalopathy treatment, mineral absorption augmentation, prebiotic action, colon carcinogenesis treatment, and inflammatory bowel disease treatment have been extensively explored. Because these lactose derivatives exhibit a variety of noteworthy physiological characteristics, they have potential applications in the food, health, and pharmaceutical sectors (Xiao et al., 2019).

The market value of lactose derivatives is typically greater than that of lactose. Ongoing research efforts are expected to result in the development of additional products in the future, which will create significant opportunities for the use of lactose derivatives as functional food ingredients. According to Data Bridge Market Research (2022), the global lactose and derivatives market was valued at USD 40.00 billion (in 2021) is expected to reach the value of USD 72.95 billion by 2029, at a CAGR of 7.80% during the forecast period. Lactulose, a commonly used laxative, was one of the initial products to achieve commercial success. Currently, there is growing interest in prebiotic lactose derivatives, such as lactosucrose or other galacto-oligosaccharides (Jelen, 2009). Several oligosaccharides are currently available for purchase, with lactulose being a well-established pharmaceutical product. Additionally, lactose and permeate powder continue to dominate the lactose industry as the most substantial components. This chapter covers the most relevant information about lactose derivatives, including its synthesis and purification, beneficial effects at physiological level, and also its potential applications.

9.2 Derivatives of Lactose

9.2.1 Lactulose

Lactulose ($C_{12}H_{22}O_{11}$; 4-O-β-D-galactopyranosyl-D-fructofuranose) is an isomer of lactose. It is a disaccharide of galactose and fructose. It does not find naturally and can be obtained by alkaline or enzymatic isomerization of lactose. Lactulose as liquid syrup is yellowish and odouless with a sweet taste while dry lactulose is a white odorless crystal, with a melting point of 168.5–170.0 °C. The relative sweetness of lactulose is 0.6 compared to sucrose. Lactulose is highly heat stable and decomposed slightly when heated to 130 °C/10 min at low pH, which makes it a suitable ingredient for food processing (Sreeranjit & Lal, 2003).

9.2.2 *Lactitol*

Lactitol ($C_{12}H_{24}O_{11}$; 4-O-β-D-galctopyranosyl-D-glucitol or β-4′ galactosylsorbitol) is a sugar alcohol produced by catalytic hydrogenation of lactose under high temperature and pressure conditions. It is a disaccharide comprising of galactose and sorbitol. Lactitol is white, sweet, odorless, crystalline non-hygroscopic solid with molecular weight of 344 dalton. Among the three crystalline forms i.e., dihydrate, monohydrate and anhydride, monohydrate form is the most common. It is very soluble in water, stable under humid and heat conditions. It has mild and clean sweet taste. The relative sweetness of lactitol is 0.3–0.42 compared to sucrose. Heat stability of lactitol is relatively high and does not cause a Maillard reaction because aldehyde group of glucose moiety is reduced to the OH group. Lactitol has low calorific value and estimated to 2–2.5 kcal/g, which is comparable to other polyols and non-digestable carbohydrates. FDA establishes a general calorific value of lactitol as 2 kcal/g, which correspond to a reduction of 48–40% with respect to sucrose.

9.2.3 *Lactosucrose*

Lactosucrose (4^G-β-D-galactosylsucrose or 4^G-β-D-lactosylfructoside) is an synthetic oligosaccharide (or trisaccharide) consisting of galactose, glucose and fructose. It is chemically *O*-β-D-galactopyranosyl-(1 → 4)-*O*-β-D-glucopyranosyl-(1 → 2)-β-D-fructofuranoside (Lee et al., 2007). The compound is obtained through transfructosylation reaction using sucrose and lactose as substrate, catalyzed by either a β-fructofuranosidase or a levansucrase. These enzymes are able to catalyse both hydrolysis and transfer reactions. However, the reaction yields are not particularly high because of occurrence of degradation of product. In 2005, lactosucrose was listed in the Foods for Specific Health Uses (FOSHU) in Japan and has many applications in food, chemical, pharmaceutical and feed applications.

Lactosucrose has a molecular weight of 504.44 dalton and obtained as a white solid with bland taste. It is highly hygroscopic in nature which forms monoclinic platelets. The specific rotation $[\alpha]_D^{20}$ of lactosucrose was determined as +59 °C (3.8 g/100 mL water) and its melting point of 181 °C (Avigad, 1957). The relative sweetness of lactosucrose is 0.3–0.6 compared to sucrose. It was proved to be non-reducing sugar on reaction with alkaline copper reagent and upon reaction with diazouracil (reagent used to identify sucrose or sucrose-containing oligosaccharide) lactosucrose formed a green colour.

9.2.4 Lactobionic Acid (LBA)

Lactobionic acid (4-O-β-D-galactosyl-D-gluconic acid) is produced by oxidation of lactose. It is composed of galactose moiety bonded to a gluconic acid through an ether-like linkage. LBA is a white powder with a molecular weight of 358.3 dalton and melting point of 128-130 °C (Alonso et al., 2013). It is readily soluble in water but poorly dissolve in organic solvents like ethyl alcohol, methyl alcohol, acetic acid, etc. It has a sweet taste with pH-reducing effects and strong mineral-complexing properties, making it suitable for applications as a food ingredient. LBA is indigestible and fermented by the intestinal flora, probably exerting prebiotic effects. Thus, energy value estimated at 2 kcal/g.

9.2.5 Galacto-Oligosaccharides (GOS)

GOS, occasionally referred to as oligogalactosyllactose, oligogalactose, oligolactose, or transgalactooligosaccharides (TOS), are classified as prebiotics due to their indigestible property (Sangwan et al., 2011). Galactose and glucose units are the building blocks of β-linked oligosaccharides (GOS), which typically have a degree of polymerization ranging from 2 to 9 and are formed at the reducing end. They are low in calories and non-cariogenic. GOS is a transparent water-soluble substance with a consistency comparable to corn syrup (Kimura et al., 2015). It can withstand high temperatures and low pH levels because to the β-glucosidic linkages. This property enables GOS to be used as an ingredient in many food products. The calorific value of GOS was calculated to be 1.73 kcal/g, which is approximately 30–50 g/100 g of digestible carbohydrates like sucrose. It is typically less sweet than sucrose, with a sweetness level ranging from 0.3 to 0.6 times that of sucrose (Panesar et al., 2018) (Fig. 9.1).

9.3 Method of Preparation of Lactose Derivatives

Lactulose can be synthesized through the isomerization of lactose by rearranging the glucose residue to the fructose molecule. Various complicated substances, alkalies, or enzymes can serve as catalysts for the isomerization of lactose into lactulose. There are two main categories for systematic approaches to lactulose production: chemical methods and enzymatic methods that involve the use of a catalyst. The earliest production of lactulose was heating a mixture of lactose and lime at 35 °C, as reported by Montgomery and Hudson (1930). A method was created using calcium hydroxide as a catalyst for converting lactose to lactulose. The process involved combining a 60% lactose solution with 0.1% calcium hydroxide at 100–102 °C for 15 min. The resulting solution was then demineralized using a

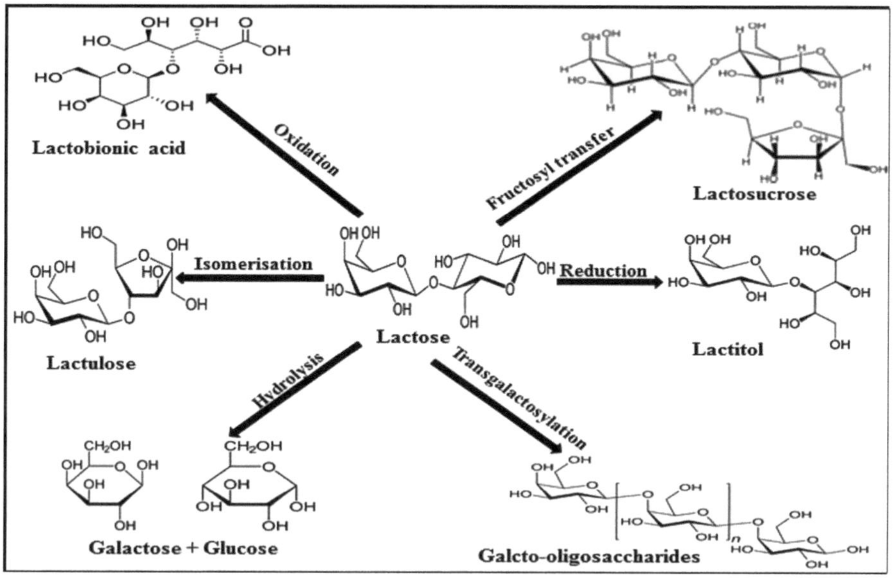

Fig. 9.1 Structure of lactose derivatives

combination of electrolysis and ion exchange resins (Matvievsky, 1979). Sodium hydroxide, magnesium oxide, and tertiary amines were utilized as alkalis.

Lactulose is enzymatically produced through the kinetically regulated transgalactosylation reaction of fructose from lactose catalyzed by β-galactosidase. In this reaction, fructose serves as the acceptor while lactose functions as both the donor and acceptor of the galactosyl group. Commercially, the primary enzymes are derived from Aspergillus and Kluyveromyces. The enzyme β-galactosidase from *Kluyveromyces lactis* is the most commonly utilized (Doles et al., 2014). It can be produced by a dual-enzymatic approach in organic-aqueous two-phase media. This involves treating a mixture of lactose and fructose with immobilized lactase and immobilized glucose isomerase (Hua et al., 2010). An enzymatic approach can address the limitations of chemical approaches in lactulose generation. Enzymatic bioconversion can lead to efficient synthesis and simple purification of the product (Panesar & Kumari, 2011).

Lactitol (4-O-β-D-Galactopyranosyl-D-glucitol) is not found in the nature and is a derivative from lactose, obtained industrially by the catalytic hydrogenation of the carbonyl group of lactose either in the presence of nickel (Ni), palladium (Pd) or ruthenium (Ru). The actual synthesis consists of a sequence of elementary reactions, namely adsorption, surface reaction, and desorption (Kevrekidis et al., 2004) and all these reactions are collectively known as the Langmuir-Hinshelwood-Hougen-Watson kinetics. The major factors which influence the lactitol formation are reaction temperature, catalyst type, agitation, and pressure of hydrogen (Schulz & Rizvi, 2023).

The first report on the synthesis of lactitol dates back to the early 1920s. The industrial production of lactitol using batch reactor is by hydrogenation of lactose in the presence of sponge nickel as a catalyst. During this reaction process, the pressure of hydrogen gas varied between 50 and 170 bar while the reaction temperature ranged from 130 to 180 °C. This type of reactor is based upon the movement of hydrogen from the gas phase to a liquid phase and across a liquid-solid interface to the surface of the supported catalyst, where the hydrogen gas is adsorbed. The advantages of batch reactors are development of negligible velocity profiles, which makes the operation simple and not having large temperature gradients.

Alternatively, production of lactitol by catalytic hydrogenation can be achieved by a continuous flow of reactants using trickle-bed reactor through structured catalysis. The reactor mainly consists of feed streams, heat exchangers, reactor units, and separator. Continuous mode of lactitol production was reported by Kasehagen (1953) using lactose solution (50% wt/wt) with a sponge nickel (1.8%) at 160 °C and 130 bar. Under such conditions, the yield of lactitol was about 98%.

Lactosucrose production by enzymatic synthesis (Levansucrase from *Aerobacter levanicum*) for first time was reported by Avigad (1957) using sucrose and lactose as fructosyl donor and acceptor, respectively as substrates. Many researchers applied enzymes like levansucrase, β-fructofuranosidase (Fujita et al., 1990) and β-galctosidase for the production of lactosucrose from lactose and sucrose as substrates. Among these levansucrase and β-fructofuranosidase are most widely studied for the production of lactosucrose. β-Fructofuranosidase is generally obtained from *Arthrobacter* sp., but for levansucrase, different microbial sources have been evaluated.

A batch reaction process was used for production of lactosucrose at an industrial level by a transfructosylation reaction using sucrose and lactose as substrates (Arakawa et al., 2002). It is produced by incubating both the substrates with the enzyme β-fructofuranosidase from *Arthrobacter* sp. and yeast (invertase-deficient) at temperature of 30-35 °C. Yeast has an important role in removing the glucose, produced as byproduct that can suppress the production of lactosucrose (Arakawa et al., 2002). This method would yield 65% lactosucrose with other saccharide mixture. The reaction involved in the production of lactosucrose as follows:

$$\text{Sucrose} + \text{Lactose} \rightarrow \text{Lactosucrose} + \text{Glucose}$$

β-Galcatosidase, is another enzyme involved in production of lactosucrose, which catalyzes the transfer of galactosyl residue from lactose to sucrose. This enzyme may cause formation of the other oligosaccharide byproducts such as GOS during the transgalactosylation (Xiao et al., 2019), which makes the purification process more difficult. Moreover, decreased yield of lactosucrose was observed since formed lactosucrose can also acts as a galactosyl donor.

Lactobionic acid production by chemical electrolytic method using lactose for first time was reported by Isbell (1933). Since then, new methods and techniques like electrochemical process (Kokoh & Alonso-Vante, 2006), the biocatalytic–microbial source (Stodola & Lockwood, 1947), the enzymatic method (Lin et al.,

1993), the catalytic hydrogenation, and heterogeneous catalytic oxidation (Kuusisto et al., 2007) were used for production of LBA. Chemical oxidation method has been mainly used in industry for production of LBA. Contrary to the chemical synthesis being an energy-intensive process that needs the use of metal catalysts, the biocatalytic approaches offer added benefits due to their high selectivity; specificity and mild reaction condition (Alonso et al., 2013). One of key strategies for the production of LBA is the use of multi-enzymes in synthesis and combining several reactions in one pot without isolating intermediates. Bi-enzymatic systems including carbohydrate oxidase/catalase (Nordkvist et al., 2007), cellobiose dehydrogenase/laccase (Yang et al., 2021) have been reported for the production of LBA.

Galactooligosaccharides are synthesized through transgalactosylation catalyzed by β-galactosidase using lactose as both the glycosyl-acceptor and -donor. During the transgalactosylation procedure, galactose units are combined to create GOS with different levels of polymerization, ultimately resulting in a glucose end unit (Mahoney, 1998). The general mechanism for producing GOS is depicted in the following equation.

$$\begin{aligned} &\mathit{Enzyme} + \mathit{lactose} \rightarrow \text{Complex}\left(\text{Enzyme} - \text{lactose}\right) \rightarrow \text{Galactosyl} - \text{enzyme} \\ &+\text{Glucose} \rightarrow \text{Galactosyl} - \text{enzyme} + \text{Acceptor} \rightarrow \text{Galactosyl} - \text{acceptor} + \text{Enzyme} \end{aligned}$$

Under hydrolysis reaction conditions, free galactose is produced when water acts as the acceptor. GOS occurs when the acceptor is a sugar. Two enzymes that can be utilized in the production of GOS are glycosyltransferases and glycohydrolases. β-galactosidases from fungi such as Aspergillus and yeasts like Kluyveromyces, Rhodotorula, *Bullera singularis*, and Sterigmatomyces, as well as bacteria from the genera Lactobacillus and Bacillus, are commonly employed as biocatalysts in the industrial production of GOS. Bacillus circulans ß-galactosidase has been chosen by producers because it can handle higher temperatures and produce more GOS than other commercial enzymes. These enzymes are well-known for their glycoside hydrolase activity, which results in the breakdown of β-galactosides into monosaccharides (Otieno, 2010).

High substrate concentrations and reduced water activity are crucial for producing galactooligosaccharides (Ambrogi et al., 2023). Furthermore, elevated temperature enhances the process of GOS production by boosting substrate solubility, allowing for the utilization of enzymes from thermophilic organisms (Nakkharat & Haltrich, 2006). β-Galactosidase enzyme exhibits low stability at reaction temperatures beyond 50 °C. Researchers are currently utilizing enzyme immobilization and protein engineering/genetic engineering to improve the stability of β-galactosidase under extreme conditions encountered in industrial biological processes such as extreme pH and temperature, low water activity, and non-natural substrates. GOS are traditionally manufactured in batch-mode using stirred tank reactors (STR). This process involves deactivating the biocatalyst and then separating it to produce GOS products without enzymes (Martins et al., 2019). A different approach to batch processes is using ultrafiltration (UF)-assisted biocatalytic reactors, sometimes referred to as enzymatic membrane reactors (EMR). Continuous glucose oxidase production in EMRs occurs in a stirred tank reactor with a free or sometimes immobilized and suspended enzyme, connected to an external membrane module. The membrane traps the enzyme in the reaction vessel, allowing saccharide fractions with smaller molecular weights to flow through. Continuous synthesis of GOS and separation of biocatalyst can be done simultaneously (Cao et al., 2020).

9.4 Application of Lactose Derivatives

Lactulose, lactitol, lactosucrose, LBA and GOS are commercially produced for a variety of food and pharmaceutical applications. Following sections discussed the applications of various lactose derivatives.

9.4.1 Lactulose

Lactulose was approved by the FDA in the United States in 1977, although there are concerns regarding the lack of data to support its efficacy. It was used to improve the sensory, browning, and sweetness characteristics of several food products such as cake, cookies, chocolate, confectionery, chewing gum, yoghurt, and ice cream (Scholz-Ahrens et al., 2001). Lactulose incorporation influence increase in probiotic counts, acidifying rate, and, as a result, milk acidity. Thus, the pH value in lactulose-containing samples was reduced towards the end of cold storage. Lactulose has been utilized as a protective agent to help some probiotics survive. In this context, the survival of *L. rhamnosus* in the presence of lactulose was examined. Lactulose-containing samples contained more than 10–60% of the strain than control samples, indicating that lactulose can improve probiotic survivability in the human colon. Researchers also discovered a shortened doubling time in conjugation with higher viability retention of *Bifidobacterium spp.* in functional skim milk containing lactulose after 4 weeks storage (Bruno et al., 2002). Lactulose also has a safeguarding impact on Lactobacillus bacteria against bile acids under harsh gastrointestinal circumstances and during refrigeration of the products. Lactulose can be utilized as a substitute for sucrose in confectionery products due to its capacity to enhance texturizing, coloration (browning), flavor, solubility, stabilizing properties, and viscosity, while maintaining a sweetness level of 0.6–0.8 of sucrose. Additionally, lactulose exhibits antiplaque effects and provides an energy value of 2.0 kcal/g (Schumann, 2002). It has been used into newborn formula as a bifidus component. Incorporating 0.5% lactulose is adequate to enhance the bifidobacterial flora to a level comparable to that reported in breastfed infants (Nagendra et al., 1995). One of its most useful qualities is its high thermostability under acidic circumstances, which enables its utilization in acidic food like fruit juice.

9.4.2 Lactitol

Over the last past decades, lactitol has emerged into a multipurpose ingredient in various food applications such as bakery, confectionery, chocolate, desserts, chewing gum, cryoprotectant, delivery agent, and stabilizer in biosensors (Martinez-Monteagudo et al., 2020). Lactitol has been approved as a safe product by the Joint FAO/WHO Expert Committee of Food Additives (JECFA) in 1983 and as a food additive (sweetening agent E966) by the European Union (Annex II of Regulation [EC] No 1333/2008). It has a self-affirmed GRAS status in the USA. Lactitol can be used not only as a prebiotic source but also as a low caloric sweetener, humectant, plasticizers, bulking agent and cryo-protectant. For synthesis of surfactants, emulsifiers, polymers, and hydrogels, lactitol has also been widely used. Numerous patents were declared in a variety of applications (such as animal feed, sugar-free chewing gum, surfactant, cleaning goods, and low-calorie chocolate and confectionary) (Cheng & Martínez-Monteagudo, 2019).

Lactitol has prebiotic properties in dairy products such as yoghurts and ice cream, since it promotes the proliferation of probiotic microorganisms, including various genera of lactic acid bacteria (Bartkiene et al., 2020). It forms a solid structure that traps the protein system and prevents it from unfolding. Additionally, it has the capability to establish hydrogen bonds with the protein structure, so aiding in the retention of the enzyme's functionality. In their study, Psimouli and Oreopoulou (2013) examined the potential of lactitol as a sugar replacement in cake recipes. They found that, based on sensory analysis, lactitol showed no significant differences compared to the control cake in terms of most sensory properties. Additionally, the batter made with lactitol exhibited a comparable flow index and temperature of starch gelatinization. Zoulias et al. (2002) conducted a study to examine the impact of lactitol, a substitute for sucrose, on the texture profile of cookie dough. The findings indicated that lactitol resulted in moderate levels of hardness and consistency, and relatively high levels of cohesiveness and adhesiveness. These texture attributes were similar to those observed in cookie dough formulated with sucrose. Lactitol, when used in Chhana-murki (an Indian dairy dessert), resulted in a favorable color, appearance, sweetness, and general acceptance (Gurditta et al., 2015). A reduction of 50% in the caloric content was achieved by the use of a combination of Splenda and lactitol. However, this combination had a negative impact on the viscosity and the ability to assimilate air (Santos & Silva, 2012). Santana et al. (2018) examined the impact of various cryoprotectants on surimi powder. They found that the addition of lactitol at a concentration of 5% not only prevented protein denaturation during freeze-drying, but also exhibited superior cryoprotective properties compared to other substances such as sorbitol and maltodextrin, as evidenced by physiochemical measurements including whiteness, gel formation, and foaming. Regarding the stabilizing action of lactitol, Klewicki (2007) found that only 3% of the oligosaccharides underwent hydrolysis due to the presence of lactitol. In addition, lactitol exhibited a safeguarding effect on the activity of α-amylase when subjected to heating. The addition of lactitol at 5% can enhance the stability of the isoenzyme glutathione transferase activity (Karamitros and Labrou, 2017).

9.4.3 Lactosucrose

Lactosucrose was classified has FOSHU ingredient in Japan for use as a probiotic dietary supplement and it is largely used in diverse healthy foods and drinks to modify the gastrointestinal conditions and improve well-being. It can be included in foods and beverages such as ice creams, infant formula, bakery products, snacks, cookies, desserts or candies (Naito et al., 1994; Ueda et al., 2000; Takumi et al., 2001) as a low-cariogenic sweetener. In several patents it has been claimed for its incorporation in chocolates, chewing gum, instant juice, instant soup and mineral water. Because of its increased water holding capacity, lactosucrose is added in fermented foods to replace the fat and improve consistence and texture properties (Silvério et al., 2015). Interestingly, compared to other lactose-based prebiotics, it has greater laxative threshold and the daily intake recommended is around 3 g.

9.4.4 Lactobionic Acid

It is a potential ingredient of the food industry due to the numerous benefits and properties as antioxidant, antimicrobial, texturizer, moisturizer, and acidulant agent. Lactobionate is a strong chelator of calcium and other divalent cations. It is commercially applied as ion sequestrant in detergent solutions, and as calcium chelator in dietary supplements. The FDA (U.S. Food and Drug Administration) considers the LBA salt (i.e., calcium lactobionate) safe for human consumption while used in sufficient amount for the desired effect on the food (Food and Drug Administration, 2018). LBA can be used as a firming agent in gels or to improve the dispersion effect and reduce hardening in pudding mixtures (Gutiérrez et al., 2012). LBA showed to be an efficient acidifier in dairy products that need a gelation step (Ribeiro et al., 2016). The sensory attributes of the cheeses made using LBA as an ingredient are considered satisfactory and fully suitable, according to a patented (U.S. Patent No. 7329424 B2, 2001) process for manufacturing cheese products and other dairy products using LBA (Koka et al., 2001). LBA also presents potential in the reduction of souring time and preservation of aroma, including the power to mask off-flavors especially in yoghurt and cheese production. Goderska (2019) reported that LBA can be employed as a direct stimulant for *bifidobacteria* development in culture and food products. The antibacterial activity of LBA against *Pseudomonas fluorescens* (Wojciechowska et al., 2020) and methicillin-resistant *Staphylococcus aureus* has also been reported (Kang et al., 2020).

9.4.5 Galactooligosaccharides

Galactooligosaccharides (GOS) is considered generally recognized as safe (GRAS) in the United States, has a non-Novel Food status in the EU, and is recognized as FOSHU in Japan. It has desirable properties that make them suitable for incorporation into a wide range of food products. These properties include their capacity to withstand variations in pH and temperature, pleasant flavor, moderate sweetness, and low caloric value. Historically, GOS have been frequently utilized as food ingredients in certain nations for an extended period of time (Tzortzis & Vulevic, 2009). Due to their acid and temperature resistance, high solubility, low glycemic index, and negligible carcinogenic properties, they are well-suited for use in the confectionery sector with reduced sugar content, as well as in hard candies with reduced calorie content (Grosová et al., 2008).

GOS are primarily utilized in fermented dairy goods such as yoghurt, as well as nutrition bars, health drinks, beverages, morning cereals, bakery items, meat products, mineral supplements, weight loss products, green foods, soups and sauces, pet food, and infant food. Their prebiotic properties rendered them suitable for nutritional and functional purposes, such as texture modification, fat substitution, and moisture preservation (Figueroa-González et al., 2011). Consequently, their incorporation into a wide range of food products, including chewing gum, confectionery, yoghurt, ice cream, and bakery infant formulas, has been steadily growing.

GOS has a wide range of pharmacological uses. Due to their similarity to human milk oligosaccharides, they were incorporated into growing-up milk and newborn formula at a concentration of 2.4 g/L. This is done to stimulate the growth of bifidobacteria and lactobacilli in the infant's intestine. They are employed in the formulation of tailored nutrition for geriatric individuals and patients in medical facilities (Lamsal, 2012). A low quantity of GOS in infant formula (0.24 g/100 mL) can encourage the growth of intestinal bifidobacteria and lactobacilli. Because of its great solubility, GOS can be easily added to dairy applications such as yoghurts, buttermilk, and dairy-based drinks. GOS also has the potential to create bread and baked items that are high in fiber, low in sugar, and low in calories. It combines beneficial features such as minimal calories and high moisture holding capacity, making it a suitable component for baked products.

9.5 Health Benefits of Lactose Derivatives

Most of the lactose derivatives viz. lactulose, lactitol, lactosucrose and GOS possess prebiotic property i.e. they provide health benefits to the host which are mediated by metabolites of microbial fermentation. For many years' lactulose is added in infant formula as bifidus factor and also used to treat constipation, hepatic encephalopathy and tooth decay while lactobionic acid possesses antioxidant activity by chelating the iron and also used in preservation solutions for the storage of organs. The following sections described the health benefits of the lactose derivatives.

9.5.1 Lactulose

Lactulose, a prebiotic sugar, may travel the upper gastrointestinal tract (comprising the stomach and small intestine) without undergoing hydrolysis, owing to the non-hydrolysable β-glycosidic link it contains. Bifidobacterium and Lactobacillus species in the colon metabolize lactulose, resulting in the production of carbon dioxide gas, acetic acid, lactic acid, and formic acid. Lactulose can function as a laxative by attracting fluid into the colon, which in turn softens the stool. This is due to the acids that are produced when lactulose is metabolized (Panesar & Kumari, 2011). Lactulose undergoes bacterial fermentation in the colon, resulting in the degradation of lactulose into SCFAs. As a result, acidifies intestinal content, causing an elevation in osmotic pressure and an increase in the volume of feces, which leads to a drop in pH. The drop in pH promotes the production of ammonium ions, which, unlike ammonia, are unable to pass through the intestinal mucosa by diffusion. This can be characterized by the laxative effect of lactulose, which results in a shorter duration for the transportation of waste through the colon and an acceleration of bowel movements. These variables are conducive to maintaining a healthy gut. Nevertheless, unlike other frequently employed osmotic laxatives, lactulose has the potential to induce electrolyte imbalances in susceptible persons (Williams et al., 2015).

It promotes the growth of specific beneficial bacteria, such as Lactobacilli and Bifidobacteria, in the colon of both adults and newborns. The health-enhancing properties of lactulose make it a perfect component for the creation of functional foods aimed at increasing or sustaining general well-being (Olano & Corzo, 2009). The impact of lactulose on both bacteria and bacterial endotoxin in the colon establishes a theoretical foundation for its application in the management of inflammatory bowel disease. It can potentially cause the following effects: alteration of the bacteria in the colon, resulting in a decrease in harmful microorganisms (Davis et al., 2007); limitation of infections by reducing the occurrence of urinary and respiratory tract infections; decrease in the production and absorption of toxic substances in the gut; regulation of bowel movements, particularly beneficial for individuals with constipation. Thus, lactulose could serve as an alternate therapeutic option for managing inflammatory bowel disease (Chen et al., 2011).

9.5.2 Lactitol

There has been a surge of research in recent years on the physiological role of lactitol. According to Natah et al. (1997), only a little amount, specifically 2%, of lactitol was absorbed and digested in the small intestine through passive diffusion. The remainder of lactitol, however, entered the colon without undergoing digestion. Lactitol underwent gradual fermentation within the large intestine, serving as an energy source for the microorganisms present in the colon. Subsequently, it underwent a conversion process resulting in the production of short-chain fatty acids

(SCFAs), lactic acid, butyric acid, carbon dioxide (CO_2), and a small amount of hydrogen (H_2). Consequently, lactitol has been demonstrated to be a significant prebiotic that promotes the growth of Lactobacilli and Bifidobacteria (Finney et al., 2007). The combination of lactitol (prebiotics) and *Lactobacillus acidophilus* NCFM (probiotics) showed mutually positive effects on the makeup of the microbiota and the functions of the mucosal lining. A recent study showed that consuming a combination of *Bifidobacterium bifidum* and *Lactobacillus acidophilus*, along with lactitol, was effective in eliminating OXA-48-producing Enterobacteriaceae bacteria. The quantification of this metabolite is employed as a preventive approach to inhibit the occurrence of intestinal translocations in neutropenic individuals and to avert the development of pneumonia (Ramos-Ramos et al., 2020).

Lactitol is highly tolerable compared to other sugar alcohols. Miller et al. (2014) found no adverse laxative effects when the daily intake was below 20 g/day. When lactitol is introduced into the mouth and undergoes fermentation, it can help prevent tooth damage. They performed a meta-analysis on the efficacy and tolerance of lactitol for adult constipation. Study based on meta-analysis on the efficacy and tolerance of lactitol for adult constipation and it was found that lactitol supplementation not only was well tolerated but also significantly improves symptoms of constipation.

Lactitol stimulates peristalsis, resulting in an increase in fecal volume. Studies conducted on the efficacy of lactitol in treating various forms of hepatic encephalopathy in infants, children, and elderly individuals have shown favorable outcomes. Lactitol is commonly prescribed as the initial treatment for hepatic encephalopathy due to its ability to decrease the absorption and generation of ammonia, as well as lower the intestinal pH (Goh et al., 2018). Recent advancements in nanomedicine have resulted in the emergence of poly-lactitol as a versatile carrier for liver cancer therapy (Hong et al., 2019). An increase in Lactobacillus and Bifidobacteria was seen, along with a potential stabilizing effect on *Blautia coccoides-Eubacterium rectale* and Clostridium cluster XIV levels, as reported by Björklund et al. (2012). Despite being classified as "GRAS" by the FDA, excessive use of lactose-derived prebiotics can lead to negative consequences such as osmotic diarrhea, abdominal pain, and vomiting (Nath et al., 2018). According to Oku et al. (2005), the highest amount of lactitol that may be taken by Japanese adults without causing temporary diarrhea is 0.36 g/kg of body weight.

9.5.3 Lactosucrose

Consumption of lactosucrose has been reported for several physiological benefits. These benefits especially result from its resistance to digestion and ability to be fermented by intestinal microbiota. Lactosucrose acts as prebiotic and is a putative growth stimulator of intestinal bifidobacteria (Arakawa et al., 2002). On microbial fermentation of lactosucrose produces SCFAs as end products. SCFAs have recognized beneficial effect to the host by reducing the intestinal pH, inhibits the proliferation of pathogenic bacteria and enhancing mineral bioavailability (Mu et al., 2013). Both *in vivo* and *in vitro* studies with human and animals has been demonstrated the intestinal fermentation of lactosucrose as well as it positive health effects. In addition, the consumption of lactosucrose has also been associated to other potential beneficial effects such as enhancement of amino acid metabolism, immunoregulatory function (Hino et al., 2007), reduces body fat accumulation (Mizote et al., 2009) prevention of colitis and inflammatory bowel disease (Zhou et al., 2015) and anti-allergic activity (Taniguchi et al., 2007). Additionally, Kishino et al. (2015) reported administration of lactoscurose suppresses the infection by influenza A virus in mice. Recently Ido et al. (2017) suggested the potential effect of lactosucrose on the improvement of psychotic symptoms of schizophrenia and improved bowel movements in schizophrenia patient.

9.5.4 Lactobionic Acid

It has significant medical interest due to the antioxidant potential and used for skin treatments because of its anti-aging capacity. In pharmaceutical applications, lactobionate most prominently used in preservation solutions (at 100 mmol/L) for storage of transplant organs since, iron chelating properties of lactobionate prevent formation of reactive oxygen species and reduce oxidative tissue injury during storage. Lactobionate is cell impermeant and thus prevents hypothermically induced cell swelling. An *in vivo* (rats) study of LBA for its antiobesity efficiency was performed by Mukherjee and Yun (2015) and results showed a significant reduction in the lipogenic capacity for both oral administration and intraperitoneal injection of the acid. There was considerable decline of the body weight gain and improvement in metabolic parameters, which makes it a potential antiobesity compound.

9.5.5 Galacto-Oligosaccharides

In addition to the benefits that GOS provide to food manufacturers through their physicochemical properties, GOS also have several health-promoting properties. This is primarily because they can regulate the balance of the microbiota in the colon and stimulate the growth of bifidobacteria in the intestines (Tzortzis & Vulevic, 2009). GOS offers multiple health advantages through two mechanisms. Firstly, it selectively promotes the growth of good bacteria, particularly Bifidobacteria and Lactobacilli in the gut. These bacteria help prevent the colonization of pathogens, hence lowering both external and internal intestinal infections. These advantageous microorganisms regulate the immune system and inhibit inflammation associated with inflammatory bowel disease (IBD). Another mechanism involves the production of SCFAs. The breakdown of GOS results in the synthesis of SCFAs. These SCFAs have diverse advantageous effects such as lowering the risk of cancer, enhancing mineral absorption, promoting regular bowel movements, regulating serum lipid and cholesterol levels, and mitigating the risk of cancer and inflammation associated with inflammatory bowel disease (IBD) (Bouhnik et al., 1997). It is challenging to measure the rise in the production of SCFAs in the colon of humans due to the quick absorption of SCFAs in the gut. Therefore, the quantity of SCFAs in the feces may not accurately reflect their production within the colon (Cherbut, 2003).

The laxative effect of GOS is attributed to its function as a soluble fiber. Oligosaccharides are not broken down during digestion and instead enter the large intestine where they promote the growth of bacteria and fermentation in the colon. Bacterial fermentation of oligosaccharides promotes bacterial growth, leading to an increase in fecal volume. Undigested oligosaccharides and fermentation products can have an osmotic impact in the gut, leading to an increase in the water content of feces. It can modify bowel function by inducing changes in the colonic environment. GOS also has a significant impact on reducing lactose sensitivity and mitigating constipation. A clinical investigation conducted by Teuri and Korpela (1998) showed that the administration of GOS to women suffering from constipation for a duration of 3 weeks resulted in a considerable alleviation of constipation symptoms.

GOS has low sweetness, reduced caloric content, and is resistant to conversion into cholesterol and fat. Consequently, it can enhance lipid metabolism, decrease serum total cholesterol levels, elevate the proportion of high-density lipoprotein in the bloodstream, and thereby successfully prevent and treat hypertension and hyperlipidemia. The study conducted by Hashmi et al. (2016) discovered a substantial decrease in serum total cholesterol, LDL, and triglycerides in rats that were fed GOS, as compared to the control group.

9.6 Commercially Available Lactose Derivatives

Lactulose is marketed as syrup and in the form of anhydrous crystals. Lactitol is commercially produced through catalytic hydrogenation of lactose. It is marked as crystalline powder available either as monohydrate or in anhydrous form. LBA is available in crystalline white powder produced by chemical or electrochemical oxidation of lactose. Lactosucrose is marked as white solid powder or syrup produced through enzymatic synthesis using lactose and sucrose as substrate. GOS is obtained as concentrated syrup or solid powder through transgalactosylation catalyzed by β-galactosidase using lactose. Table 9.1 shows few major producers of lactose derivatives and its forms.

Table 9.1 Major producers of lactose derivatives and its forms

Lactose derivatives	Major manufacturing companies	Forms of available in market
Lactobionic acid	Global Calcium (India) Merck & Co., Inc. (USA) Avantor, Inc. (USA), Reliable Biopharmaceutical Corp. (United States) Novartis AG (Switzerland) Dayang Chem (Hangzhou) Co., Ltd. Solvay (Germany) Sandoz (Germany) Friesland Campina domo (the Netherlands)	White powder in crystalline form
Lactosucrose	Ensuiko Sugar Refining Co. (Japan) Daiwa Kasei Co. Ltd. (Japan) Hayashibara Shoji Inc. (Japan)	Powder (45–70% on dry matter) or as syrup containing 72% to 78% (w/v) of solid content of the lactosucrose
Lactitol	Mitsubishi Shoji Foodtech Co (Japan) Purac Biochem (The Netherlands) DuPont Danisco (Denmark) B-Food Science (Japan)	Concentrated syrup or crystalline powder in anhydrous and monohydrate form
Lactulose	Henan Tianfu Chemical Co., Ltd (Hong Kong) Solactis (France) Fresenius Kabi (Austria and Italy Inalco (Italy) DaniPharm (Denmark) Dephasaar Chem.-Pharm. Fabrik GmbH (Germany) Morinaga Milk Industry Co. (Japan and Germany)	Syrup or anhydrous crystalline powder
Galacto-oligosaccharides	Baolingbao Biology Co. Ltd (China). Clasado BioSciences (UK), Fonterra Co-operative Group Ltd (New Zealand).	Concentrated syrup or solid powder

9.7 Conclusions

Lactose is a principal constituent carbohydrate of mammalian milk. Lactose derivatives can be obtained from lactose using various methods, such as oxidation, reduction, isomerization and enzymatic polymerization. Because of high market value of lactose derivatives viz. lactulose, lactosucrose, lactitol, and GOS offers significant opportunities for value-added conversion of lactose to functional food ingredients. Lactose derivatives have various effects and many of them have been put to practical use including infant products and confectionaries. Because of prebiotic effects, low sweetening power and low glycemic index of lactose derivatives demands from the pharmaceutical and dietary supplement industries surge the market growth.

References

Aït Aissa, A., & Aïder, M. (2014). Electro-catalytic isomerization of lactose into lactulose: The impact of the electric current, temperature and reactor configuration. *International Dairy Journal, 34*(2), 213–219. https://doi.org/10.1016/j.idairyj.2013.08.010

Alonso, S., Rendueles, M., & Díaz, M. (2013). Bio-production of lactobionic acid: Current status, applications and future prospects. *Biotechnology Advances, 31*(8), 1275–1291. https://doi.org/10.1016/j.biotechadv.2013.04.010

Ambrogi, V., Bottacini, F., Cao, L., Kuipers, B., Schoterman, M., & van Sinderen, D. (2023). Galacto-oligosaccharides as infant prebiotics: Production, application, bioactive activities and future perspectives. *Critical Reviews in Food Science and Nutrition, 63*(6), 753–766. https://doi.org/10.1080/10408398.2021.1953437

Arakawa, K., Aoyama, Y., Ikeda, H., Mikuni, K., Fujita, K., & Hara, K. (2002). The development of Lactosucrose production and its applications in foods for specified health use. *Journal of Applied Glycoscience, 49*(1), 63–72. https://doi.org/10.5458/jag.49.63

Avigad, G. (1957). Enzymatic synthesis and characterization of a new trisaccharide, alpha-lactosyl-beta-fructofuranoside. *The Journal of Biological Chemistry, 229*(1), 121–129. https://doi.org/10.1016/s0021-9258(18)70600-5

Bartkiene, E., Lele, V., Ruzauskas, M., Domig, K. J., Starkute, V., Zavistanaviciute, P., Bartkevics, V., Pugajeva, I., Klupsaite, D., Juodeikiene, G., Mickiene, R., & Rocha, J. M. (2020). Lactic acid bacteria isolation from spontaneous sourdough and their characterization including antimicrobial and antifungal properties evaluation. *Microorganisms, 8*(1), 64. https://doi.org/10.3390/microorganisms8010064

Björklund, M., Ouwehand, A. C., Forssten, S. D., Nikkilä, J., Tiihonen, K., Rautonen, N., & Lahtinen, S. J. (2012). Gut microbiota of healthy elderly NSAID users is selectively modified with the administration of lactobacillus acidophilus NCFM and lactitol. *Age, 34*(4), 987–999. https://doi.org/10.1007/s11357-011-9294-5

Bouhnik, Y., Flourié, B., D'Agay-Abensour, L., Pochart, P., Gramet, G., Durand, M., & Rambaud, J. C. (1997). Administration of transgalacto-oligosaccharides increases fecal bifidobacteria and modifies colonic fermentation metabolism in healthy humans. *Journal of Nutrition, 127*(3), 444–448. https://doi.org/10.1093/jn/127.3.444

Bruno, F. A., Lankaputhra, W. E. V., & Shah, N. P. (2002). Growth, viability and activity of Bifidobacterium spp. in skim milk containing prebiotics. *Journal of Food Science, 67*(7), 2740–2744. https://doi.org/10.1111/j.1365-2621.2002.tb08807.x

Cao, T., Pázmándi, M., Galambos, I., & Kovács, Z. (2020). Continuous production of galacto-oligosaccharides by an enzyme membrane reactor utilizing free enzymes. *Membranes, 10*(9), 1–16. https://doi.org/10.3390/membranes10090203

Chen, X., Zuo, Q., Hai, Y., & Sun, X. J. (2011). Lactulose: An indirect antioxidant ameliorating inflammatory bowel disease by increasing hydrogen production. *Medical Hypotheses, 76*(3), 325–327. https://doi.org/10.1016/j.mehy.2010.09.026

Chen, Q., Xiao, Y., Zhang, W., Zhang, T., Jiang, B., Stressler, T., Fischer, L., & Mu, W. (2018). Current research on cellobiose 2-epimerase: Enzymatic properties, mechanistic insights, and potential applications in the dairy industry. *Trends in Food Science and Technology, 82*, 167–176. https://doi.org/10.1016/j.tifs.2018.09.009

Cheng, S., & Martínez-Monteagudo, S. I. (2019). Hydrogenation of lactose for the production of lactitol. *Asia-Pacific Journal of Chemical Engineering, 14*(1), 2275. https://doi.org/10.1002/apj.2275
Cherbut, C. (2003). Motor effects of short-chain fatty acids and lactate in the gastrointestinal tract. *Proceedings of the Nutrition Society, 62*(1), 95–99. https://doi.org/10.1079/pns2002213
Data Bridge Market Research. (2022). Global antimicrobials market – Industry trends and forecast to 2029. https://www.databridgemarketresearch.com/reports/global-lactose-and-derivatives-market.
Davis, P. F., Greenhill, N. S., Rowan, A. M., & Schollum, L. M. (2007). The safety of New Zealand bovine colostrum: Nutritional and physiological evaluation in rats. *Food and Chemical Toxicology, 45*(2), 229–236. https://doi.org/10.1016/j.fct.2006.07.034
Doles, P. E., Babu, S. M., Ittera, N., John, I., John, L., & John, N. (2014). Tapioca starch based antimicrobial food packaging material. *Ijret, 3*(9), 1051–1056.
Figueroa-González, I., Quijano, G., Ramírez, G., & Cruz-Guerrero, A. (2011). Probiotics and prebiotics-perspectives and challenges. *Journal of the Science of Food and Agriculture, 91*(8), 1341–1348. https://doi.org/10.1002/jsfa.4367
Finney, M., Smullen, J., Foster, H. A., Brokx, S., & Storey, D. M. (2007). Effects of low doses of lactitol on faecal microflora, pH, short chain fatty acids and gastrointestinal symptomology. *European Journal of Nutrition, 46*(6), 307–314. https://doi.org/10.1007/s00394-007-0666-7
Food and Drug Administration (FDA). (2018). Title 21 – Food and drugs. Chapter I – Part 172. Food additives permitted for direct addition to food for human consumption act 2011. Pub, *L*(172),72. https://www.accessdata.fda.gov/scripts/cdrh/cfdocs/cfcfr/CFRSearch.cfm?fr=172.720
Fujita, K., Hara, K., Hashimoto, H., & Kitahata, S. (1990). Purification and some properties of β-fructofuranosidase I from Arthrobacter sp. K-1. *Agricultural and Biological Chemistry, 54*(4), 913–919.
Ganzle, M. G., Haase, G., & Jelen, P. (2008). Lactose: Crystallization, hydrolysis and value-added derivatives. *International Dairy Journal, 18*(7), 685–694.
Goderska, K. (2019). The antioxidant and prebiotic properties of lactobionic acid. *Applied Microbiology and Biotechnology, 103*(9), 3737–3751. https://doi.org/10.1007/s00253-019-09754-7
Goh, E. T., Stokes, C. S., Sidhu, S. S., Vilstrup, H., Gluud, L. L., & Morgan, M. Y. (2018). L-ornithine L-aspartate for prevention and treatment of hepatic encephalopathy in people with cirrhosis. *Cochrane Database of Systematic Reviews, 5*, CD012410.
Grosová, Z., Rosenberg, M., Rebroš, M., Šipocz, M., & Sedláčková, B. (2008). Entrapment of β-galactosidase in polyvinylalcohol hydrogel. *Biotechnology Letters, 30*(4), 763–767. https://doi.org/10.1007/s10529-007-9606-0
Gurditta, H., Patel, A. A., & Arora, S. (2015). Optimisation of sweetener and bulking agent levels for the preparation of functional Chhana-murki. *International Journal of Dairy Technology, 68*(2), 190–197. https://doi.org/10.1111/1471-0307.12175
Gutiérrez, L. F., Hamoudi, S., & Belkacemi, K. (2012). Lactobionic acid: A high value-added lactose derivative for food and pharmaceutical applications. *International Dairy Journal, 26*(2), 103–111. https://doi.org/10.1016/j.idairyj.2012.05.003
Hashmi, A., Naeem, N., Farooq, Z., Masood, S., Iqbal, S., & Naseer, R. (2016). Effect of prebiotic Galacto-oligosaccharides on serum lipid profile of Hypercholesterolemics. *Probiotics and Antimicrobial Proteins, 8*(1), 19–30. https://doi.org/10.1007/s12602-016-9206-1
Hino, K., Kurose, M., Sakurai, T., Inoue, S., Oku, K., Chaen, H., & Fukuda, Y. (2007). Effect of dietary Lactosucrose (4G-. BETA.-D-Galactosylsucrose) on the intestinal immune functions in mice. *Journal of Applied Glycoscience, 54*(3), 169–172.
Hong, S. J., Ahn, M. H., Sangshetti, J., & Arote, R. B. (2019). Sugar alcohol-based polymeric gene carriers: Synthesis, properties and gene therapy applications. *Acta Biomaterialia, 97*, 105–115. https://doi.org/10.1016/j.actbio.2019.07.029
Hua, X., Yang, R., Zhang, W., Fei, Y., Jin, Z., & Jiang, B. (2010). Dual-enzymatic synthesis of lactulose in organic-aqueous two-phase media. *Food Research International, 43*(3), 716–722. https://doi.org/10.1016/j.foodres.2009.11.008

Ido, Y., Nagamine, T., Okamura, T., Tasaki, M., & Fukuo, K. (2017). Prebiotic lactosucrose may improve not only constipation but also psychotic symptoms of schizophrenia. *International Medical Journal, 24*(3), 305–306.

Isbell, H. S. (1933). Preparation of calcium lactobionate and lactobionic delta-lactone. *Bureau of Standards Journal of Research, 11*(5), 713. https://doi.org/10.6028/jres.011.048

Jelen, P. (2009). Dried whey, whey proteins, lactose and lactose derivative products. In *Dairy powders and concentrated products* (pp. 255–267). https://doi.org/10.1002/9781444322729.ch7

Kang, S., Kong, F., Shi, X., Han, H., Li, M., Guan, B., Yang, M., Cao, X., Tao, D., Zheng, Y., & Yue, X. (2020). Antibacterial activity and mechanism of lactobionic acid against Pseudomonas fluorescens and methicillin-resistant Staphylococcus aureus and its application on whole milk. *Food Control, 108*, 106876. https://doi.org/10.1016/j.foodcont.2019.106876

Karamitros, C. S., & Labrou, N. E. (2017). Preserving enzymatic activity and enhancing biochemical stability of glutathione transferase by soluble additives under free and tethered conditions. *Biotechnology and Applied Biochemistry, 64*(5), 754–764. https://doi.org/10.1002/bab.1535

Kasehagen, L. (1953). *Continuous hydrogenation of sugars U.S.: Patent and trademark office*. Patent and Trademark Office.

Kevrekidis, I., Shvartsman, Y., & S. (2004). Chemical reactor analysis and design fundamentals. *Chemical Engineering Science, 59*(10), 2123–2124. https://doi.org/10.1016/j.ces.2003.12.030

Kimura, L. J., McGee, A., Baird, S., Viloria, J., & Nagatsuka, M. (2015). Healthy mothers healthy babies: Awareness and perceptions of existing breastfeeding and postpartum depression support among parents and perinatal health care providers in Hawai'i. *Hawai'i Journal of Medicine & Public Health : A Journal of Asia Pacific Medicine & Public Health, 74*(3), 101–111.

Kishino, E., Takemura, N., Masaki, H., Ito, T., & Nakazawa, M. (2015). Dietary lactosucrose suppresses influenza A (H1N1) virus infection in mice. *Bioscience of Microbiota, Food and Health, 34*(4), 67–76. https://doi.org/10.12938/BMFH.2015-005

Klewicki, R. (2007). Effect of selected parameters of lactose hydrolysis in the presence of β-galactosidase from various sources on the synthesis of galactosyl-polyol derivatives. *Engineering in Life Sciences, 7*(3), 268–274. https://doi.org/10.1002/elsc.200620185

Koka, R., Mehnert, D. W., Fritsch, R. J., Steffan, W., Habermeier, P., Bradbury, A. G. W., Alan Wolfschoon-Pombo, M. R., & Gitte Budolfsen Lynglev, H. P. H.-H. (2001). Process for manufacturing cheese and other dairy products and products thereof. *Patent, 2*(12), 8.

Kokoh, K. B., & Alonso-Vante, N. (2006). Electrocatalytic oxidation of lactose on gold nanoparticle modified carbon in carbonate buffer. *Journal of Applied Electrochemistry, 36*(2), 147–151. https://doi.org/10.1007/s10800-005-9042-6

Kuusisto, J., Tokarev, A. V., Murzina, E. V., Roslund, M. U., Mikkola, J. P., Murzin, D. Y., & Salmi, T. (2007). From renewable raw materials to high value-added fine chemicals-catalytic hydrogenation and oxidation of d-lactose. *Catalysis Today, 121*(1–2), 92–99. https://doi.org/10.1016/j.cattod.2006.11.020

Lamsal, B. P. (2012). Production, health aspects and potential food uses of dairy prebiotic galactooligosaccharides. *Journal of the Science of Food and Agriculture, 92*(10), 2020–2028. https://doi.org/10.1002/jsfa.5712

Lee, J. H., Lim, J. S., Park, C., Kang, S. W., Shin, H. Y., Park, S. W., & Kim, S. W. (2007). Continuous production of lactosucrose by immobilized Sterigmatomyces elviae mutant. *Journal of Microbiology and Biotechnology, 17*(9), 1533–1537.

Lin, S. F., Hu, H. M., Inukal, T., & Tsai, Y. C. (1993). Production of novel oligosaccharide oxidase by wheat bran solid-state fermentation. *Biotechnology Advances, 11*(3), 417–427. https://doi.org/10.1016/0734-9750(93)90011-B

Lu, L., Guo, L., Wang, K., Liu, Y., & Xiao, M. (2020). β-Galactosidases: A great tool for synthesizing galactose-containing carbohydrates. *Biotechnology Advances, 39*, 107465. https://doi.org/10.1016/j.biotechadv.2019.107465

Mahoney, R. R. (1998). Galactosyl-oligosaccharide formation during lactose hydrolysis: A review. *Food Chemistry, 63*(2), 147–154. https://doi.org/10.1016/S0308-8146(98)00020-X

Martínez-Monteagudo, S. I., Enteshari, M., & Metzger, L. (2019). Lactitol: Production, properties, and applications. *Trends in Food Science & Technology, 83*, 181–191.
Martinez-Monteagudo, S., Rathnakumar, K., Enteshari, M., Nyuydze, C., Osorio-Arias, J. C., & Ranaweera, H. (2020). Hundred years of Lactitol: From hydrogenation to food ingredient. *Lactose and Lactose Derivatives*. https://doi.org/10.5772/intechopen.93365
Martins, G. N., Ureta, M. M., Tymczyszyn, E. E., Castilho, P. C., & Gomez-Zavaglia, A. (2019). Technological aspects of the production of fructo and galacto-oligosaccharides. Enzymatic synthesis and hydrolysis. *Frontiers in Nutrition, 6*, 78. https://doi.org/10.3389/fnut.2019.00078
Matvievsky, V. (1979). *Investigation of process for lactose/lactulose syrup production for infantile food*. Master of Science thesis (p. 252). Uglich Institute.
Miller, L. E., Tennilä, J., & Ouwehand, A. C. (2014). Efficacy and tolerance of lactitol supplementation for adult constipation: A systematic review and meta-analysis. *Clinical and Experimental Gastroenterology, 7*(1), 241–248. https://doi.org/10.2147/CEG.S58952
Mizote, A., Taniguchi, Y., Takei, Y., Koya-Miyata, S., Kohno, K., Iwaki, K., Kurose, M., Oku, K., Chaen, H., & Fukuda, S. (2009). Lactosucrose inhibits body fat accumulation in rats by decreasing intestinal lipid absorption. *Bioscience, Biotechnology and Biochemistry, 73*(3), 582–587. https://doi.org/10.1271/bbb.80658
Montgomery, E. M., & Hudson, C. S. (1930). Relations between rotatory power and structure in the sugar group. XXVII. Synthesis of a new disaccharide ketose (lactulose) from lactose. *Journal of the American Chemical Society, 52*(5), 2101–2106. https://doi.org/10.1021/ja01368a060
Mu, W., Chen, Q., Wang, X., Zhang, T., & Jiang, B. (2013). Current studies on physiological functions and biological production of lactosucrose. *Applied Microbiology and Biotechnology, 97*(16), 7073–7080. https://doi.org/10.1007/s00253-013-5079-3
Mukherjee, R., & Yun, J. W. (2015). Lactobionic acid reduces body weight gain in diet-induced obese rats by targeted inhibition of galectin-1. *Biochemical and Biophysical Research Communications, 463*(4), 1311–1316. https://doi.org/10.1016/j.bbrc.2015.06.114
Nagendra, R., Viswanatha, S., Kumar, S. A., Murthy, B. K., & Rao, S. V. (1995). Effect of feeding milk formula containing lactulose to infants on faecal bifidobacterial flora. *Nutrition Research, 15*(1), 15–24. https://doi.org/10.1016/0271-5317(95)91649-W
Naito, F., Aoyama, Y., Fujita, K., & Takahashi, S. (1994). The characteristics of lactosucrose (LS-55P) when processed in cooking (Part 1) – Effects on physical properties and taste of marshmallow. *Science of Cookery, 27*(4), 253–259.
Nakkharat, P., & Haltrich, D. (2006). Purification and characterisation of an intracellular enzyme with β-glucosidase and β-galactosidase activity from the thermophilic fungus Talaromyces thermophilus CBS 236.58. *Journal of Biotechnology, 123*(3), 304–313. https://doi.org/10.1016/j.jbiotec.2005.12.015
Natah, S. S., Hussien, K. R., Tuominen, J. A., & Koivisto, V. A. (1997). Metabolic response to lactitol and xylitol in healthy men. *American Journal of Clinical Nutrition, 65*(4), 947–950. https://doi.org/10.1093/ajcn/65.4.947
Nath, A., Haktanirlar, G., Varga, Á., Molnár, M. A., Albert, K., Galambos, I., Koris, A., & Vatai, G. (2018). Biological activities of lactose-derived prebiotics and symbiotic with probiotics on gastrointestinal system. *Medicina (Lithuania), 54*(2), 18. https://doi.org/10.3390/medicina54020018
Nordkvist, M., Nielsen, P. M., & Villadsen, J. (2007). Oxidation of lactose to lactobionic acid by a Microdochium nivale carbohydrate oxidase: Kinetics and operational stability. *Biotechnology and Bioengineering, 97*(4), 694–707. https://doi.org/10.1002/bit.21273
Oku, T., Nakamura, S., & Ichinose, M. (2005). Maximum permissive dosage of lactose and lactitol for transitory diarrhea, and utilizable capacity for lactose in Japanese female adults. *Journal of Nutritional Science and Vitaminology, 51*(2), 51–57. https://doi.org/10.3177/jnsv.51.51
Olano, A., & Corzo, N. (2009). Lactulose as a food ingredient. *Journal of the Science of Food and Agriculture, 89*(12), 1987–1990.
Otieno, D. O. (2010). Synthesis of β-Galactooligosaccharides from lactose using microbial β-galactosidases. *Comprehensive Reviews in Food Science and Food Safety, 9*(5), 471–482. https://doi.org/10.1111/j.1541-4337.2010.00121.x

Panesar, P. S., & Kumari, S. (2011). Lactulose: Production, purification and potential applications. *Biotechnology Advances, 29*(6), 940–948. https://doi.org/10.1016/j.biotechadv.2011.08.008

Panesar, P. S., Kaur, R., Singh, R. S., & Kennedy, J. F. (2018). Biocatalytic strategies in the production of galacto-oligosaccharides and its global status. *International Journal of Biological Macromolecules, 111*, 667–679. https://doi.org/10.1016/j.ijbiomac.2018.01.062

Psimouli, V., & Oreopoulou, V. (2013). The effect of fat replacers on batter and cake properties. *Journal of Food Science, 78*(10), 99–105. https://doi.org/10.1111/1750-3841.12235

Ramos-Ramos, J. C., Lázaro-Perona, F., Arribas, J. R., García-Rodríguez, J., Mingorance, J., Ruiz-Carrascoso, G., Borobia, A. M., Paño-Pardo, J. R., Herruzo, R., & Arnalich, F. (2020). Proof-of-concept trial of the combination of lactitol with Bifidobacterium bifidum and lactobacillus acidophilus for the eradication of intestinal OXA-48-producing Enterobacteriaceae. *Gut Pathogens, 12*(1), 1–8. https://doi.org/10.1186/s13099-020-00354-9

Ribeiro, J. C. B., Granato, D., Masson, M. L., Andriot, I., Mosca, A. C., Salles, C., & Guichard, E. (2016). Effect of lactobionic acid on the acidification, rheological properties and aroma release of dairy gels. *Food Chemistry, 207*(1), 101–106. https://doi.org/10.1016/j.foodchem.2016.03.066

Sangwan, V., Tomar, S. K., Singh, R. R. B., Singh, A. K., & Ali, B. (2011). Galactooligosaccharides: Novel components of designer foods. *Journal of Food Science, 76*(4), 103–111. https://doi.org/10.1111/j.1750-3841.2011.02131.x

Santana, P., Zilda, D. S., & Huda, N. (2018). Physicochemical properties of surimi powder made from threadfin bream (*Nemipterus japonicus*) with various dryoprotectants added. *Journal of Fundamental and Applied Sciences, 9*(2S), 866. https://doi.org/10.4314/jfas.v9i2s.54

Santos, G. G., & Silva, M. R. (2012). Sorvete de mangaba (Hancornia speciosa Gomez) preparado com substitutos de gordura e açúcar. *Ciência e Tecnologia de Alimentos, 32*(3), 621–628. https://doi.org/10.1590/S0101-20612012005000086

Scholz-Ahrens, K. E., Schaafsma, G., van den Heuvel, E. G., & Schrezenmeir, J. (2001). Effects of prebiotics on mineral metabolism. *The American Journal of Clinical Nutrition, 73*(2), 459s–464s.

Schulz, P., & Rizvi, S. S. H. (2023). Hydrolysis of lactose in Milk: Current status and future products. *Food Reviews International, 39*(5), 2875–2894. https://doi.org/10.1080/87559129.2021.1983590

Schumann, C. (2002). Medical, nutritional and technological properties of lactulose. An update. *European Journal of Nutrition, 41*(SUPPL. 1), 17–25. https://doi.org/10.1007/s00394-002-1103-6

Silvério, S. C., Macedo, E. A., Teixeira, J. A., & Rodrigues, L. R. (2015). Perspectives on the biotechnological production and potential applications of lactosucrose: A review. *Journal of Functional Foods, 19*, 74–90. https://doi.org/10.1016/j.jff.2015.09.014

Sreeranjit, C. V. K., & Lal, J. J. (2003). GLUCOSE | properties and analysis. In *Encyclopedia of food sciences and nutrition* (pp. 2898–2903). https://doi.org/10.1016/b0-12-227055-x/00557-5

Stodola, F. H., & Lockwood, L. B. (1947). The oxidation of lactose and maltose to bionic acids by pseudomonas. *Journal of Biological Chemistry, 171*(1), 213–221. https://doi.org/10.1016/s0021-9258(17)41119-7

Takumi, H., Ochi, H., Okada, S., Li, S.-T., Terada, A., & Mitsuoka, T. (2001). Effect of ingesting frozen yoghurt in combination with Lactosucrose consumption on the fecal microbiota and fecal Metabolitic activity in healthy adults. *Japanese Journal of Food Microbiology, 18*(2), 49–56. https://doi.org/10.5803/jsfm.18.49

Taniguchi, Y., Mizote, A., Kohno, K., Iwaki, K., Oku, K., Chaen, H., & Fukuda, S. (2007). Effects of dietary lactosucrose (4G-β-D-galactosylsucrose) on the IgE response in mice. *Bioscience, Biotechnology and Biochemistry, 71*(11), 2766–2773. https://doi.org/10.1271/bbb.70364

Teuri, U., & Korpela, R. (1998). Galacto-oligosaccharides relieve constipation in elderly people. *Annals of Nutrition and Metabolism, 42*(6), 319–327. https://doi.org/10.1159/000012751

Tzortzis, G., & Vulevic, J. (2009). Galacto-Oligosaccharide Prebiotics. In D. Charalampopoulos & R. A. Rastall (Eds.), *Prebiotics and probiotics science and technology* (pp. 207–244). Springer.

Ueda, K., Beppu, H., Maruyama, M., Sakaki, T., Shirasuna, F., Inoue, T., Kawai, K., Tamai, I., Ikeda, H., Fujita, K., & Kuzuya, H. (2000). Development of oligosaccharide-containing croissant. *Journal of the Japanese Society for Food Science and Technology, 47*(11), 836–843.

Williams, C. F., Walton, G. E., Jiang, L., Plummer, S., Garaiova, I., & Gibson, G. R. (2015). Comparative analysis of intestinal tract models. *Annual Review of Food Science and Technology, 6*, 329–350. https://doi.org/10.1146/annurev-food-022814-015429

Wojciechowska, A., Klewicki, R., & Klewicka, E. (2020). The potential of new bionic acids as prebiotics and antimicrobials. *Lwt, 125*, 109246. https://doi.org/10.1016/j.lwt.2020.109246

Xiao, Y., Chen, Q., Guang, C., Zhang, W., & Mu, W. (2019). An overview on biological production of functional lactose derivatives. *Applied Microbiology and Biotechnology, 103*, 3683–3691. https://doi.org/10.1007/s00253-019-09755-6

Yang, J., Xu, P., Long, L., & Ding, S. (2021). Production of lactobionic acid using an immobilized cellobiose dehydrogenase/laccase system on magnetic chitosan spheres. *Process Biochemistry, 100*, 1–9. https://doi.org/10.1016/j.procbio.2020.09.024

Zhou, Y., Ruan, Z., Zhou, X., Huang, X., Li, H., Wang, L., Zhang, C., Deng, Z., Wu, G., & Yin, Y. (2015). Lactosucrose attenuates intestinal inflammation by promoting Th2 cytokine production and enhancing CD86 expression in colitic rats. *Bioscience, Biotechnology and Biochemistry, 79*(4), 643–651. https://doi.org/10.1080/09168451.2014.991680

Zoulias, E. I., Oreopoulou, V., & Kounalaki, E. (2002). Effect of fat and sugar replacement on cookie properties. *Journal of the Science of Food and Agriculture, 82*(14), 1637–1644. https://doi.org/10.1002/jsfa.1230

Chapter 10
Challenges and Opportunities in Processing of Lactose Hydrolyzed Dairy Products

Sonu K. Shivanna and Priyanka Singh Rao

10.1 Introduction

The processing of lactose hydrolyzed dairy products presents a dynamic landscape marked by both challenges and opportunities. Lactose intolerance, a common condition affecting a significant portion of the global population, necessitates the development of dairy products with reduced lactose content or completely devoid of it. This has spurred the innovation and refinement of lactose hydrolysis techniques, aimed at breaking down lactose into its constituent sugars, glucose, and galactose, thereby rendering dairy products more digestible for lactose-intolerant individuals. However, the journey from raw dairy to lactose-hydrolyzed products is fraught with challenges. The enzymatic or chemical hydrolysis processes employed can be complex, requiring precise control of conditions such as temperature, pH, and enzyme activity to achieve optimal lactose breakdown without compromising product quality. For instance, the preparation of lactose-hydrolyzed milk involves various thermal processing steps, which accounts for lower acceptance of these products (Troise et al., 2016). Maillard browning exacerbates during subsequent storage, leading to deterioration in quality, shelf life, and nutritional value of the product (Singh et al., 2021). Furthermore, maintaining the nutritional integrity, taste, and texture of the final dairy products post-hydrolysis presents a significant challenge that must be addressed through careful formulation and processing techniques.

Despite these challenges, the processing of lactose-hydrolyzed dairy products presents promising opportunities for both consumers and the dairy industry. For consumers, access to a wider range of dairy options, including milk, yogurt, cheese,

S. K. Shivanna (✉)
Dairy Chemistry Division, ICAR-National Dairy Research Institute, Karnal, Haryana, India

P. S. Rao
Dairy Chemistry Section, Southern Regional Station, ICAR-National Dairy Research Institute, Bengaluru, Karnataka, India

P. S. Rao et al. (eds.), *Lactose Hydrolysis in Dairy Products*,
https://doi.org/10.1007/978-3-031-78207-7_10

and ice cream, without the discomfort associated with lactose intolerance, enhances dietary inclusivity and satisfaction. Additionally, the burgeoning market for lactose-free and low-lactose dairy products reflects a growing consumer demand for healthier and more easily digestible alternatives. For the dairy industry, the production of lactose-hydrolyzed products represents a strategic avenue for innovation and differentiation in an increasingly competitive market. By leveraging advanced processing technologies and ingredient formulations, dairy manufacturers can capitalize on the rising consumer interest in lactose-free and low-lactose options, thereby expanding their product portfolios and tapping into new market segments.

In this context, this chapter explores the multifaceted landscape of processing lactose-hydrolyzed dairy products, examining the challenges inherent in lactose hydrolysis techniques, as well as the opportunities presented by the growing demand for lactose-free alternatives. Through a comprehensive analysis of industry trends, technological advancements, and consumer preferences, we aim to elucidate key strategies for optimizing the production and commercialization of lactose-hydrolyzed dairy products in the modern food market.

10.2 Challenges in Processing of Lactose Hydrolyzed Dairy Products

At present, commercially available milk is typically manufactured using either enzymatic batch hydrolysis or a combination of enzymatic batch hydrolysis and membrane filtration techniques to create lactose-free milk products with higher protein content and equivalent sweetness to regular milk. Comparatively, enzymatic hydrolysis offers more benefits over filtration, with only a marginal increase in milk price, estimated at $0.06 to $0.08 (Schulz & Rizvi, 2023). Transitioning from batch processing to continuous processing would bring economic and environmental advantages, as it allows for enzyme recovery, thus further reducing the cost of enzymatic hydrolysis. However, the implementation of continuous enzymatic hydrolysis on a commercial scale has been hindered by challenges such as scaling up batch reactors and the recovery and cleaning of immobilized enzymes. Research focusing on the theoretical scaling up of enzymatic bioreactors is crucial for advancing the commercialization of this process. Additionally, the recent surge in patents related to the immobilization of enzymes like ß-galactosidase indicates promising progress. Further exploration into the feasibility and creation of a large-scale bioreactor utilizing immobilized ß-galactosidase that can be efficiently reused is warranted.

10.2.1 Sensory and Quality Issues

Due to the significant influence of sensory attributes on consumer preferences, manufacturers of lactose-free dairy products place considerable emphasis on sensory characteristics. Lactose-free dairy products resembling skim milk are often perceived as more desirable. However, conventional lactose-free milk available in the market typically employs lactase to break down lactose into galactose and glucose, resulting in a sweeter taste compared to regular milk. Additionally, commercial lactase may exhibit proteolytic activity, leading to the release of peptides and free amino acids, which in turn can contribute to non-enzymatic browning and undesirable flavors during storage. The degree of proteolysis and subsequent product deterioration is contingent upon the type of lactase preparation utilized and its purity (Troise et al., 2016). Notably, there is limited research exploring the impact of microbial sources on proteolytic activity.

Nielsen et al. (2018) conducted a study comparing five commercial lactase preparations of varying purities from different companies. They found that lactase preparations with lower levels of purification exhibited greater degrees of proteolysis and aggregation. Consequently, lactose-free milk became more prone to Maillard reactions, with storage leading to a significant increase in hexosylation, up to 11-fold, in lactose-free UHT milk (Milkovska-Stamenova & Hoffmann, 2017). Lactose-free milk is often characterized by a cooked and eggy flavor, which recent research attributes to volatile sulfur compounds generated from Maillard reactions between lactose and amino acids such as cysteine and methionine (Jo et al., 2018).

In batch processing of lactose-free milk, heat treatment typically eliminates lactase and its proteolytic activity. In contrast, in an aseptic process, lactase remains active throughout the product's shelf life. This sustained activity of lactase during storage could potentially compromise the quality of the product (Tossavainen & Kallioinen, 2007).

10.2.2 Challenges During Drying and Storage

10.2.2.1 Impact of Lactose Hydrolysis on Drying Process

The hydrolysis of lactose is known to release low molecular weight monosaccharides, namely glucose and galactose, both of which exhibit reduced glass transition temperatures (Tg) compared to lactose (Shrestha et al., 2007). This temperature directly impacts the drying process, making it more challenging. In fact, Schuck et al. (2016) noted that adjustments and adaptations to drying parameters are necessary for each equipment, production, and product during the drying process to achieve uniformity in the final product. Consequently, the food industry may face difficulties such as product adhesion, equipment clogging, low yield, browning, and crystallization during both drying and storage (Fialho et al., 2018; Torres et al., 2017; Shrestha et al., 2007; Roos, 2002). Hence, it is important to recognize that employing the same drying operational parameters used for conventional powdered milk is not feasible for powdered milk with hydrolyzed lactose.

10.2.2.2 Chemical Alterations During Storage

Producing shelf-stable Ultra-high temperature hydrolyzed-lactose milk (UHLM) presents a challenge for dairy manufacturers, as the product undergoes chemical alterations during storage caused by the reactivity of reducing sugars and proteolysis stemming from impurities in the lactase preparations. The presence of proteolytic side activity in the lactase preparation (LP) results in an increased concentration of free amino groups in the milk, thereby accelerating the formation rate of volatile Maillard Reaction Products (MRPs) (Jansson et al., 2014). In addition to addressing proteolytic side activity, it was found that arylsulfatase side activity in lactose preparation could lead to undesirable off-flavors during storage, resulting from the formation of p-cresol from naturally occurring sulphonated-cresol in the milk (Dekker et al., 2019). To prevent shelf-life issues, producers should use high-quality lactases, such as arylsulfatase-free lactases available in Maxilact® products. Meanwhile, the stability of UHLM during export/import is also a major concern since the transportation process, marked by temperature variations and stringent quarantine protocols during customs inspections, poses frequent challenges when shipping products to foreign destinations. In such conditions, Maillard reactions occur extensively in UHLM, heightening the risk of quality degradation before reaching consumers. While extending the shelf-life of UHLM to match that of regular UHT milk (9–12 months) would facilitate distribution, addressing the critical factors limiting UHLM shelf-life necessitates a comprehensive understanding of the production technology and the associated chemical modifications involved.

10.2.2.3 Maillard Reaction

During storage, dairy products face significant challenges, particularly concerning the Maillard reaction, a reaction between reducing sugars like lactose and lysine-rich proteins, leading to the formation of various MRPs. Lactose hydrolyzed milk (LHM) and lactose hydrolyzed milk powder (LHMP) are especially prone to maillard browning due to the higher reactivity of monosaccharides (glucose and galactose) compared to lactose with lysine. However, these changes, including browning and off-flavors, stemming from high heat treatments, can compromise product quality. Moreover, β-galactosidases used in lactose hydrolysis may exhibit proteolytic activity, releasing peptides and free amino acids during storage, resulting in elevated levels of strecker aldehydes and increased MRP formation (Troise et al., 2016). On the other hand, maillard reaction progresses more extensively in Ultra-high temperature hydrolyzed-lactose milk (UHLM) compared to regular UHT milk (Jansson et al., 2014). This heightened susceptibility to non-enzymatic browning is the primary reason why producers have opted to reduce the commercial shelf-life of the product to 90–120 days. In lactose-free UHT milk, Maillard reaction susceptibility increases during storage, leading to elevated hexosylation levels and a characteristic cooked and eggy flavor due to Maillard reactions between lactose and amino acids (Milkovska-Stamenova & Hoffmann, 2017; Jo et al., 2018). The presence of advanced glycation end products (AGEs), such as carboxymethyl lysine (CML) and carboxyethyl lysine (CEL), formed from Amadori products, can lead to protein modifications and induce oxidative stress and inflammation upon binding to cell receptors. Despite attempts to mitigate the Maillard reaction using enzymes like fructosamine oxidase and fructosamine kinase, hydrogen peroxide generation during their reaction and the presence of metal ions may lead to undesired protein modification and lipid peroxidation (Singh et al., 2021). Addition of polyphenols to dairy products to mitigate maillard browning may result in off-flavors and increased discoloration, albeit in small quantities compared to amines. Nevertheless, protein phenol adducts formed by polyphenols influence protein functionality in milk systems, and oxidation during heating and storage can exacerbate bitterness (Singh et al., 2021).

Furthermore, Lactose hydrolyzed milk powder exhibits greater sensitivity to changes in environmental conditions compared to traditional milk powder. This heightened sensitivity can manifest in various issues, including prolonged rehydration time, reduced solubility, browning, aggregation, and caking, as noted by several studies (Lule et al., 2016; Shrestha et al., 2007; Schuck et al., 2005; Fernandez et al., 2003). Additionally, the occurrence of high-temperature peaks during storage, regardless of their timing, has an accumulative effect on modifying powder properties (Nasser et al., 2017). Hence, controlling the transport and storage conditions of lactose hydrolyzed milk powder is crucial for preserving product quality. Research indicates that lactose hydrolyzed milk powder has limited stability when exposed to high temperatures or humidity levels, leading to storage-related technological challenges. A study evaluating the molecular structure of powders subjected to accelerated aging (at 60 °C for 8 hours and 24 hours) aims to elucidate how lactose hydrolysis contributes to these issues over time (Fialho et al., 2021). The evidence suggests that the process of aggregation facilitated by Maillard reactions unfolds in two distinct stages: initially, glycation disrupts the native arrangement and/or conformation of proteins, leading to partial unfolding of the protein structure and heightened flexibility. Subsequently, these alterations facilitate the formation of cross-links between the proteins (Fialho et al., 2021).

10.3 Sweetness of the Developed Products

The challenge in creating low lactose milk and milk products lies in reducing lactose, the natural sugar in milk, which can affect their taste. When lactose is reduced, the resulting lactose-hydrolyzed milk has a higher cryoscopic value, increasing from 0.454 to 0.650 °C, and a stronger sweet taste because lactose is less sweet than glucose and galactose (Troise et al., 2016). Additionally, lactose hydrolysis doubles the amount of reducing sugars in the milk, which promotes Maillard reaction (Tossavainen & Kallioinen, 2007). Lactose-free milk powder tends to be more hygroscopic because it contains more molecules in the amorphous state (glucose and galactose) during drying. Spray-drying of skim milk with hydrolyzed lactose results in low cyclone recovery and significant powder sticking inside the spray dryer (Shrestha et al., 2007). Higher levels of lactose hydrolysis led to increased adhesion to the drying chamber due to greater particle agglomeration (Torres et al., 2017). Moreover, increased levels of reducing sugar in hydrolyzed milk powder can cause non-enzymatic browning during high-temperature processing, resulting in more Maillard reaction products and odors. In the case of low lactose ice cream, the increase in monosaccharide content after lactose hydrolysis lowers the freezing point of the ice cream mix, resulting in a softer texture at the same temperature. While this softer texture might be desirable for some frozen desserts because of the "soft scoop," it also causes faster melting. However, since lactose hydrolysis increases sweetness, the ice cream maker can reduce sugar addition to raise the melting temperature again. Adjusting the recipe in this way may decrease the total solids in the mix, necessitating additional measures to comply with local ice cream regulations (Dekker et al., 2019).

In order to mitigate the usual doubling of sweetness following lactose hydrolysis and to restore a traditional taste, methods such as ultrafiltration, nanofiltration, or chromatography (alongside the hydrolysis of remaining lactose) are employed. This yields high-quality milk with a flavor very close to that of regular milk (Dekker et al., 2019; Jelen & Tossavainen, 2003; Harju et al., 2012).

10.4 Scaling Up Cost: Batch to Continuous

The cost of scaling up from batch to continuous production for low lactose milk can be significant due to various factors. In batch processing, lactose reduction is typically done in discrete batches, which can be time-consuming and labor-intensive. However, switching to continuous production offers several advantages, such as higher efficiency and consistency, albeit with increased initial investment. To produce lactose-free milk, the enzyme β-galactosidase can be directly added to whole milk. After reaching the desired level of lactose hydrolysis, heat treatment deactivates the enzyme. However, since the free enzyme cannot be reused, this method is not cost-effective. To address this, immobilized β-galactosidase is used for hydrolyzing skim milk. Once the desired lactose hydrolysis is achieved, cream is added to adjust the milk's fat content. Although several hydrolysis systems have been studied, few have been scaled up, and even fewer have been applied industrially. The first commercial company to hydrolyze lactose in milk using immobilized lactase was Centrale del Latte in Milan, Italy, using SNAM Progetti technology. They used immobilized Saccharomyces (Kluyveromyces) lactis lactase in cellulose triacetate fibers. Sumitomo Chemical, Japan, developed a process to immobilize fungal-origin β-galactosidase on an amphoteric ion-exchange resin made of phenol formaldehyde polymer, which was used by Drouin Cooperative Butter Factory for producing market milk and hydrolyzed whey (Honda et al., 1993). Snow Brand's factory developed a rotary column reactor, functioning as both a stirred tank reactor and a packed bed reactor (Honda et al., 1991). The packing density of immobilized β-galactosidase greatly influenced the reaction rate. This reactor addressed issues like channeling or severe pressure drops. In horizontal rotary columns, 70%–80% lactose hydrolysis was observed, with the enzyme washed for 36 cycles, indicating its suitability for milk hydrolysis. Snow Brand subsequently set up a commercial plant based on this. However, microbial contamination, protein adherence, and channeling are major challenges in immobilized enzyme systems. Periodic washing and pasteurization are crucial for long-term operations (Panesar et al., 2010). Protein adhered to the enzyme can be dissolved with high and low pH solutions. Pasteurization with benzalkonium chloride is also used. Changing the flow direction of feed can overcome channeling in packed column systems (Panesar et al., 2010). Overall, while batch to continuous scaling offers long-term benefits in terms of efficiency, consistency, and cost savings, the initial investment can be significant. Careful planning, feasibility studies, and risk assessment are essential to ensure a successful transition while managing scaling-up costs effectively.

10.5 Future Opportunities of Lactose Hydrolyzed Dairy Product Development

The potential in lactose hydrolyzed dairy product processing is supported by scientific research, indicating a growing market demand and consumer preferences. Deng et al. (2015) explored consumer preferences for lactose-free dairy products in China, revealing a strong demand among individuals with lactose intolerance. Similarly, Misselwitz et al. (2013) highlighted the global prevalence of lactose intolerance and its impact on consumer behavior, emphasizing the need for lactose-free options.

10.5.1 Potential for Product Diversification

The production of low-lactose and lactose-free milk powders has become increasingly popular in recent years. This is due to their versatility in various dairy and bakery products, as well as the added benefit of reducing logistical costs associated with powder storage, transportation, and packaging (Singh et al., 2021). The demand for lactose-free dairy products has created opportunities for product diversification within the dairy industry, with manufacturers expanding their product lines to cater to diverse consumer preferences and dietary needs. This expansion includes traditional dairy products such as milk, yogurt, and cheese, as well as innovative formulations like lactose-free ice cream, desserts, and beverages. By offering a diverse range of lactose-free products, manufacturers can appeal to a broader consumer base and capture new market segments. Lactose-free dairy products provide essential nutrients found in regular dairy products, such as calcium and vitamins, to individuals unable to digest lactose. This category has a wide and growing health appeal to consumers, with the quality and variety of lactose-free dairy products significantly increasing in recent years.

Over the past 5 years, the variety of lactose-free dairy products has expanded to include milk, yogurt, butter, cheese, and flavored milk (Li et al., 2023). Brands such as Valio in Finland, Lactalis in France, and Fair life in the US have capitalized on this trend, offering lactose-free options with enhanced nutritional profiles. In China, dairy companies have responded to the demand by increasing the production of lactose-free products, with Monmilk Xinyangdao becoming the country's first internationally certified lactose-free dairy brand in 2017 (Li et al., 2023). These efforts aim to make lactose-free products more mainstream, standardized, and healthier options for consumers worldwide.

Furthermore, various lactose-hydrolyzed milk products (LHMP) have been developed, including lactose-hydrolyzed yogurt (Popescu et al., 2022; Yamamoto et al., 2021; Karnyaczki & Csanadi, 2017; Moreira et al., 2017), ice cream (Tvorogova et al., 2023; Osmak et al., 2021; Abbasi & Saeedabadian, 2015), and Dulce de leche (Francisquini et al., 2019). The production of low-lactose (Zhang et al., 2020; Morioka et al., 2018; Torres et al., 2017) and lactose-free (Rodrigues et al., 2020; Fialho et al., 2018) milk powders has gained considerable traction in recent years due to their versatility in applications across dairy and bakery products. Additionally, these products offer the added advantage of reduced logistical costs associated with powder storage, transportation, and packaging (Fialho et al., 2019).

10.5.2 Lactose Hydrolyzed Milk Powder

Skim milk powder (SMP) is a plentiful and economical resource containing approximately 52% lactose by weight. The United States exported a substantial 665,124 metric tons of SMP in 2018 (Levitt & Beck, 2019). With a significant portion of the global population experiencing lactose intolerance, there is a considerable opportunity to produce lactose-hydrolyzed skim milk powder (LHSMP). This could offer lactose-intolerant individuals a milk option or serve as a sweeter alternative in confectionery without additional calories. Additionally, milk powders are lucrative protein sources, especially as milk powder prices decrease. Utilizing milk protein to create products such as nutritional bars can yield a markup of up to 400% (Nielsen, 2019). However, LHSMP encounters challenges. The hydrolysis of lactose into galactose and glucose increases susceptibility to Maillard browning, particularly during drying processes such as spray drying, leading to fouling and browning issues. While freeze-drying has shown success, its high-cost limits industrial feasibility (Jouppila & Roos, 1994). To address browning, increasing the production of galacto-oligosaccharides (GOS) during lactose hydrolysis is proposed. Studies indicate that GOS, along with galactose and glucose, exhibit lower browning levels compared to galactose and glucose alone. Recent developments in lactose-free skim milk powder by Valio, such as their Eila product line, demonstrate promise. Eila SWEET, produced through enzymatic hydrolysis and spray drying, and Eila PRO, which undergoes lactose filtration before hydrolysis and spray drying, provide options with reduced carbohydrates and higher protein content, indicating a bright future for lactose-hydrolyzed skim milk powders in the dairy market (Schulz & Rizvi, 2023).

10.5.3 Lactose Free UHT Milk Enriched with Prebiotic GOS

Galactooligosaccharides (GOS), renowned for their favorable impact on the intestinal bacterial microflora, are derivable from lactose found in whey, a residual product of the cheese industry, through enzymatic conversion facilitated by lactase enzymes (Feijoo et al., 2017). Recognized for their prebiotic attributes, GOS have garnered attention across industries including infant formula, infant foods, livestock feed, and pet food (Huang et al., 2020). Patents have been secured for the enzymatic lactose-to-GOS conversion process (European Patent EP3041945B1) and for *Bifidobacterium bifidum* strains capable of producing β-galactosidase enzymes essential for this conversion (US Patent US20070274955A1). Lactase enzymes, or β-galactosidases, exhibit dual functionality in transgalactosylation and hydrolysis reactions, with reaction conditions determining the predominance of each reaction type. Notably, under conditions of low lactose concentrations (<300 mM) and high-water activity (aw), hydrolysis is favored, whereas at high lactose concentrations (>980 mM) and low aw, transgalactosylation prevails, resulting in GOs production (Vera et al., 2020).

GOS, being indigestible sugars, possess prebiotic characteristics akin to human milk oligosaccharides (Ben et al., 2008). Their robust stability in varying temperature and pH environments renders them suitable for incorporation into a diverse array of food products such as infant formulae, fruit juices, bakery items, and confectionery goods (Fischer & Kleinschmidt, 2018). The global market for GOS was valued at USD 570 million in 2021 and is anticipated to reach USD 850 million by 2027 due to escalating demand for functional foods (Market Data Forecast, 2022).

In the production of GOS-enriched lactose-free UHT milk, higher reaction temperatures have been found to increase GOS synthesis through transgalactosylation. However, temperature elevation during milk processing poses microbial growth challenges. Post-process hydrolysis, where UHT milk is hermetically sealed to prevent microbial contamination and stored at room temperature, is employed to address this issue. Aseptic packaging of milk containing β-galactosidase enables storage at elevated temperatures without spoilage, accelerating GOS production through transgalactosylation. It is important to note that higher storage temperatures increase the potential for Maillard reactions in milk (Schulz & Rizvi, 2023). Regarding oligosaccharide production, there is an inverse relationship between sweetness and degree of polymerization (Ruiz-Aceituno et al., 2018). Different enzyme sources yield GOS with varying glycosidic linkages and degrees of polymerization. Typically, GOS has relatively low sweetness, approximately 30–60% that of sucrose (Torres et al., 2010). By selecting enzymes promoting higher degree of polymerization, the relative sweetness of GOS can be reduced, enabling addition of more lactose to milk before hydrolysis. This approach enhances GOS synthesis by capitalizing on the direct relationship between lactose concentration and transgalactosylation rate, achieving high GOS yield without introducing excessive sweetness to milk products.

Further research is needed on post-process hydrolysis during storage to address potential issues. Elevated temperatures may lead to Maillard reactions, resulting in increased furosine concentration, discoloration, and lysine blockage. In the context of GOS, excessive temperatures causing Maillard browning can break glycosidic linkages, reducing degree of polymerization and yielding undesirable products. Therefore, careful monitoring and control of storage conditions are essential to optimize GOS production and maintain product quality.

10.5.4 Extended Shelf-Life Dairy Products Using Lactose Derivatives

10.5.4.1 Hydrolyzed Syrup

Lactose hydrolysis breaks down lactose into its basic sugars, glucose and galactose, resulting in a syrup that is three times sweeter than lactose (Fox, 2011). These syrups find various applications in the food industry, such as improving the properties of muffin cakes in the bakery sector (Sokoti Naji et al., 2017), or serving as sugar substitutes in white pan bread (Ogunrinola et al., 1988). They have also been explored as replacements for sucrose in canned foods like beans and plums (Chandan et al., 1982), peaches and pears (Tweedie & MacBean, 1978), and frozen desserts (Rexroat & Bradley Jr., 1986). Additionally, these syrups are suitable for individuals who are lactose intolerant once they are fully hydrolyzed. However, one consideration during syrup production is the total solids content, as sugars may crystallize above 70% total solids. Chiu and Kosikowski (1985) produced a 100% hydrolyzed syrup with 60% solids and stored it successfully for 4 weeks at room temperature without crystallization issues. Samadov et al. (2019) recommend concentrating syrups to 65% total solids for optimal sweetness and storage stability.

10.5.4.2 D-Tagatose

The process of turning lactose into syrup isn't just about making glucose-galactose syrups. It's also possible to convert the galactose part into d-tagatose, a rare sugar with potential as a low-calorie sweetener because of its stability, high solubility, and sweetness (92% as sweet as sucrose) (von Rymon Lipinski, 2013). This conversion typically happens in two steps, either enzymatically or chemically. First, lactose is broken down into its individual sugars, glucose and galactose, using a β-galactosidase enzyme. Then, the d-galactose is converted into d-tagatose, usually using the enzyme l-arabinose isomerase (Cervantes et al., 2020; Jayamuthunagai et al., 2017).

10.5.4.3 Lactulose

Lactulose, an isomer of lactose, undergoes a chemical transformation known as the Lobry de Bruyn–Alberda van Ekenstein process under alkaline conditions, converting the glucose aldehyde to a ketone (fructose). This compound, not broken down in the small intestine, reaches the colon unchanged. There, it acts as a prebiotic, promoting the growth of beneficial bacteria such as Lactobacilli and Bifidobacteria (Tuohy et al., 2002). Because of this property, lactulose is commonly used in functional foods like frozen yogurt (Awad et al., 2014). The increasing demand for such products is expected to drive the global lactulose market to a value of USD 218 million by 2028 (Research Reports World, 2022). Enzymatic lactulose synthesis involves using common glycosyltransferase and glycosidase enzymes like β-galactosidase to catalyze the transfer of galactose from lactose to fructose (Olano & Corzo, 2009). Adamczak et al. (2009) found that the efficiency of this process in whey permeate depends on factors such as the choice of enzyme and the ratio of lactose to fructose. However, a drawback of using permeate as a substrate is the absence of fructose, necessitating its supplementation (Adamczak et al., 2009). Lorenzen et al. (2013) addressed this by combining β-galactosidase with glucose isomerase, which converts glucose to fructose. With this dual enzyme approach, they successfully produced 20% galactooligosaccharides (GOS), including 1% lactulose, from ultrafiltered skim milk permeate with an initial lactose content of around 40%.

10.5.4.4 Lactobionic Acid

Lactobionic acid (LBA), chemically known as 4-O-β-galactopyranosyl-d-gluconic acid, is an organic compound consisting of a galactose molecule linked to gluconic acid through an ether-like bond. Widely used in biomedical, cosmetics, and chemical sectors, LBA also shows potential applications in the food industry, including fortification of beverages with minerals like calcium, improving flavor in fruit preparations, enhancing water retention in meat products, and preventing staling in bread (Cardoso et al., 2019). Additionally, LBA exhibits prebiotic (Goderska, 2019) characteristics and can act as a food antimicrobial (Sáez-Orviz et al., 2022). However, its toxicity hasn't been fully investigated, and thus, it's not yet regulated for food use in the EU (Cardoso et al., 2019). LBA can be produced from lactose using various methods, including chemical synthesis and bioconversion methods like enzyme or microbial catalysis. Recent research has focused on bioconversion methods due to their higher selectivity and milder conditions (Chidar, 2022). Chidar (2022) examined factors affecting lactose and whey protein (WP) conversion into LBA using different enzyme combinations and novel activation methods such as ultrasonication, microwave, and mechano-milling. Ultrasonication didn't improve LBA production, but microwave assistance significantly reduced reaction times. However, enzymatic oxidation of lactose produces hydrogen peroxide as a side product, requiring catalases to break it down.

10.6 Conclusion

In conclusion, the processing of lactose hydrolyzed dairy products presents multifaceted challenges, from enzymatic specificity and effectiveness to sensory quality issues and storage stability. However, these challenges are accompanied by significant opportunities, fueled by market demand and the potential for product diversification. Addressing challenges such as enzyme stability and optimizing processing methods are crucial for enhancing the quality and marketability of lactose-free or low lactose dairy products. Moreover, exploring innovative product formulations, such as lactose-free UHT milk enriched with prebiotic GOS, can drive growth and expansion in the lactose hydrolyzed dairy product market. Overall, continued research and development efforts are essential for capitalizing on the opportunities in this evolving sector and meeting the diverse needs of consumers seeking health-oriented dairy alternatives.

References

Abbasi, S., & Saeedabadian, A. (2015). Influences of lactose hydrolysis of milk and sugar reduction on some physical properties of ice cream. *Journal of Food Science and Technology, 52*(1), 367–374.

Adamczak, M., Charubin, D., & Bednarski, W. (2009). Influence of reaction medium composition on enzymatic synthesis of galactooligosaccharides and lactulose from lactose concentrates prepared from whey permeate. *Chemical Papers, 63*(2), 111–116.

Awad, R. A., Hagrass, A. E., Salama, W. M., Elmalek, F. A., & Eldardiry, A. I. (2014). Lactulose production from milk permeate and its performance in healthy functional frozen yoghurt. *World Journal of Dairy & Food Sciences, 9*(1), 1–9.

Ben, X. M., Li, J., Feng, Z. T., Shi, S. Y., Lu, Y. D., Chen, R., & Zhou, X. Y. (2008). Low level of galacto-oligosaccharide in infant formula stimulates growth of intestinal Bifidobacteria and Lactobacilli. *World Journal of Gastroenterology, 14*(42), 6564–6568.

Cardoso, T., Marques, C., Dagostin, J. L. A., & Masson, M. L. (2019). Lactobionic acid as a potential food ingredient: recent studies and applications. *Journal of Food Science, 84*(7), 1672–1681.

Cervantes, F. V., Neifar, S., Merdzo, Z., Viña-Gonzalez, J., Fernandez-Arrojo, L., Ballesteros, A. O., Fernandez-Lobato, M., Bejar, S., & Plou, F. J. (2020). A three-step process for the bioconversion of whey permeate into a glucose-free D-tagatose syrup. *Catalysts, 10*(6), 647.

Chandan, R. C., Uebersax, M. A., & Saylock, M. J. (1982). Utilization of cheese whey permeate in canned beans and plums. *Journal of Food Science, 47*(5), 1649–1653.

Chidar, E. (2022). *Bioconversion process assisted by non-conventional techniques for the production of lactobionic acid from lactose and dairy-by-products* (MSc thesis). McGill University. https://escholarship.mcgill.ca/concern/theses/p8418t225

Chiu, C. P., & Kosikowski, F. V. (1985). Hydrolyzed lactose syrup from concentrated sweet whey permeates. *Journal of Dairy Science, 68*(1), 16–22.
Dekker, P. J., Koenders, D., & Bruins, M. J. (2019). Lactose-free dairy products: Market developments, production, nutrition and health benefits. *Nutrients, 11*(3), 551.
Deng, Y., Misselwitz, B., Dai, N., & Fox, M. (2015). Lactose intolerance in adults: biological mechanism and dietary management. *Nutrients, 7*(9), 8020–8035.
European Patent. EP3041945B1. Production of galacto-oligosaccharides. Retrieved from https://patents.google.com/patent/EP3041945B1/en.
Feijoo, S., González-García, S., Lema, J. M., & Moreira, M. T. (2017). Life cycle assessment of β-Galactosidase enzyme production. *Journal of Cleaner Production, 165*, 204–212.
Fernandez, E., Schebor, C., & Chirife, J. (2003). Glass transition temperature of regular and lactose hydrolyzed milk powders. *LWT-Food Science and Technology, 36*(5), 547–551.
Fialho, T. L., Martins, E., Silveira, A. C. P., Silva, C. R. D. J., Perrone, Í. T., Schuck, P., & de Carvalho, A. F. (2018). Lactose hydrolyzed milk powder: Thermodynamic characterization of the drying process. *Drying Technology, 36*(8), 922–931.
Fialho, T. L., Nogueira, M. H., Moreau, A., Delaplace, G., Schuck, P., Perrone, Í. T., & Júnior, P. P. D. S. P. (2019). Sugar type matters in spray drying: Homogeneous distribution in milk powder favors repulsive interactions between proteins. *Food Structure, 22*, 100132.
Fialho, T. L., Nascimento, L. G. L., Moreau, A., Delaplace, G., Martins, E., Perrone, Í. T., & Junior, P. P. D. S. P. (2021). Sugar type matters in spray drying II: Glycation effects on physicochemical characteristics of aged lactose-hydrolyzed milk powder. *Food Structure, 30*, 100215.
Fischer, C., & Kleinschmidt, T. (2018). Synthesis of galactooligosac-charides in milk and whey: A review. *Comprehensive Reviews in Food Science and Food Safety, 17*(3), 678–697.
Fox, P. F. (2011). Lactose and oligosaccharides. In J. W. Fuquay, P. F. Fox, & P. L. H. McSweeney (Eds.), *Encyclopedia of dairy sciences* (pp. 173–181). Elsevier.
Francisquini, J. D. A., Rocha, J., Martins, E., Stephani, R., da Silva, P. H. F., Renhe, I. R. T., et al. (2019). 5-Hydroxymethylfurfural formation and color change in lactose-hydrolyzed Dulce de leche. *Journal of Dairy Research, 86*(4), 477–482.
Goderska, K. (2019). The antioxidant and prebiotic properties of lactobionic acid. *Applied Microbiology and Biotechnology, 103*(9), 3737–3751.
Harju, M., Kallioinen, H., & Tossavainen, O. (2012). Lactose hydrolysis and other conversions in dairy products: Technological aspects. *International Dairy Journal, 22*(2), 104–109.
Honda, Y., Hashiba, H., Ahiko, K., & Takahashi, H. (1991). Hydrolysis of lactose by immobilized β-galactosidase using a horizontal rotary column reactor. *Nippon Shokuhin Kogyo Gakkaishi, 38*(5), 384–390.
Honda, Y., Kako, M., Abiko, K., & Sogo, Y. (1993). Hydrolysis of lactose in milk. *Bioprocess Technology, 16*, 209–209.
Huang, J., Zhu, S., Zhao, L., Chen, L., Du, M., Zhang, C., & Yang, S. T. (2020). A novel β-galactosidase from Klebsiella oxytoca ZJUH1705 for efficient production of galacto-oligosaccharides from lactose. *Applied Microbiology and Biotechnology, 104*, 6161–6172.
Jansson, T., Clausen, M. R., Sundekilde, U. K., Eggers, N., Nyegaard, S., Larsen, L. B., & Bertram, H. C. (2014). Lactose-hydrolyzed milk is more prone to chemical changes during storage than conventional ultra-high-temperature (UHT) milk. *Journal of Agricultural and Food Chemistry, 62*(31), 7886–7896.
Jayamuthunagai, J., Srisowmeya, G., Chakravarthy, M., & Gautam, P. (2017). D-Tagatose production by permeabilized and immobilized Lactobacillus plantarum using whey permeate. *Bioresource Technology, 235*, 250–255.
Jelen, P., & Tossavainen, O. (2003). Low lactose and lactose-free milk and dairy products-prospects, technologies and applications. *Australian Journal of Dairy Technology, 58*(2), 161.
Jo, Y., Benoist, D. M., Barbano, D. M., & Drake, M. A. (2018). Flavor and flavor chemistry differences among milks processed by high-temperature, short-time pasteurization or ultra-pasteurization. *Journal of Dairy Science, 101*(5), 3812–3828.

Jouppila, K., & Roos, Y. H. (1994). Glass Transitions and Crystallization in Milk Powders. *Journal of Dairy Science, 77*(10), 2907–2915.

Karnyaczki, Z., & Csanadi, J. (2017). Texture profile properties, sensory evaluation, and susceptibility to syneresis of yoghurt prepared from lactose-free milk. *Acta Alimentaria, 46*(4), 403–410.

Levitt, A., & Beck, M. (2019). *Market outlook (2017–2019)*. US Dairy Export Council.

Li, A., Zheng, J., Han, X., Yang, S., Cheng, S., Zhao, J., & Lu, Y. (2023). Advances in low-lactose/lactose-free dairy products and their production. *Food, 12*(13), 2553.

Lorenzen, P. C., Breiter, J., Clawin-Rädecker, I., & Dau, A. (2013). A novel bi-enzymatic system for lactose conversion. *International Journal of Food Science & Technology, 48*(7), 1396–1403.

Lule, V. K., Garg, S., Tomar, S. K., Khedkar, C. D., & Nalage, D. N. (2016). Food intolerance: lactose intolerance. In B. Caballero, P. M. Finglas, & F. Toldrá (Eds.), *Encyclopedia of food and health* (pp. 43–48). Academic.

Market Data Forecast. (2022). *Galacto-oligosaccharides (GOS) market*. https://www.marketdataforecast.com/market-reports/galacto-oligosaccharide-market

Milkovska-Stamenova, S., & Hoffmann, R. (2017). Influence of storage and heating on protein glycation levels of processed lactose-free and regular bovine milk products. *Food Chemistry, 221*, 489–495.

Misselwitz, B., Pohl, D., Frühauf, H., Fried, M., Vavricka, S. R., & Fox, M. (2013). Lactose malabsorption and intolerance: pathogenesis, diagnosis and treatment. *United European Gastroenterology Journal, 1*(3), 151–159.

Moreira, T. C., da Silva, A. T., Fagundes, C., Ferreira, S. M. R., Cândido, L. M. B., Passos, M., & Krüger, C. C. H. (2017). Elaboration of yogurt with reduced level of lactose added of carob (Ceratonia siliqua L.). *LWT-Food Science and Technology, 76*, 326–329.

Morioka, L. R. I., Koga, E. C., & Suguimoto, H. H. (2018). Development and evaluation of freeze-dried milk powder with low-lactose content. *Acta Alimentaria, 47*(3), 283–290.

Nasser, S., Moreau, A., Jeantet, R., Hédoux, A., & Delaplace, G. (2017). Influence of storage conditions on the functional properties of micellar casein powder. *Food and Bioproducts Processing, 106*, 181–192.

Nielsen. Product Insider, (2019). https://www.nielsen.com/us/en/insights/report/2019/total-consumer-report-2019

Nielsen, S. D., Zhao, D., Le, T. T., Rauh, V., Sørensen, J., Andersen, H. J., & Larsen, L. B. (2018). Proteolytic side-activity of lactase preparations. *International Dairy Journal, 78*, 159–168.

Ogunrinola, O. A., Jeon, I. J., & Ponte, J. G., Jr. (1988). Functional properties of hydrolyzed whey permeate syrups in bread formulations. *Journal of Food Science, 53*(1), 215–217.

Olano, A., & Corzo, N. (2009). Lactulose as a food ingredient. *Journal of the Science of Food and Agriculture, 89*(12), 1987–1990.

Osmak, T., Mleko, S., Bass, O., Mykhalevych, A., & Kuzmyk, U. (2021). Enzymatic hydrolysis of lactose in concentrates of reconstituted demineralized whey, intended for ice cream production. *Cabi Digital Library, 10*, 277–288.

Panesar, P. S., Kumari, S., & Panesar, R. (2010). Potential applications of immobilized β-galactosidase in food processing industries. *Enzyme Research, 2010*, 473137.

Popescu, L., Bulgaru, V., & Siminiuc, R. (2022). Effects of lactose hydrolysis and milk type on the quality of lactose-free yoghurt. *Journal of Engineering Sciences, 4*, 164–175.

Research Reports World. (2022). Lactulose market report 2022–2028. *Market Watch.*. https://www.marketwatch.com/press-release/lactulose-market-size-global-2022-share-growth-segments-revenue-manufacturers-and-2028-forecast-research-report-2022-08-24

Rexroat, T. M., & Bradley, R. L., Jr. (1986). Acceptance of frozen desserts made with concentrated, decolorized, deionized hydrolyzed whey permeate. *Journal of Dairy Science, 69*(5), 1225–1231.

Rodrigues, L., Paiva, G., Lisboa, H. M., Pasquali, M., Gusmão, R., Duarte, M. E., & Abrantes, T. (2020). Impact of spray drying parameters on lactose-free milk powder properties and composition. *Journal of Agricultural Studies, 8*(3), 32.

Roos, Y. H. (2002). Importance of glass transition and water activity to spray drying and stability of dairy powders. *Le Lait, 82*(4), 475–484.

Ruiz-Aceituno, L., Hernandez-Hernandez, O., Kolida, S., Moreno, F. J., & Methven, L. (2018). Sweetness and sensory properties of commercial and novel oligosaccharides of prebiotic potential. *LWT-Food Science and Technology, 97*, 476–482.

Sáez-Orviz, S., Marcet, I., Rendueles, M., & Díaz, M. (2022). The antimicrobial and bioactive properties of lactobionic acid. *Journal of the Science of Food and Agriculture, 102*(9), 3495–3502.

Samadov, R., Ciprovica, I., Zolnere, K., & & Cinkmanis, I. (2019, May 2–3). The optimization of acid whey permeate hydrolysis for glucose-galactose syrup production. *Proceedings of the 13th Baltic conference on food science and technology*, FOODBALT 2019 "Food. Nutrition. Well-Being". (pp. 254–257).

Schuck, P., Blanchard, E., Dolivet, A., Méjean, S., Onillon, E., & Jeantet, R. (2005). Water activity and glass transition in dairy ingredients. *Le Lait, 85*(4–5), 295–304.

Schuck, P., Jeantet, R., Bhandari, B., Chen, X. D., Perrone, Í. T., de Carvalho, A. F., & Kelly, P. (2016). Recent advances in spray drying relevant to the dairy industry: A comprehensive critical review. *Drying Technology, 34*(15), 1773–1790.

Schulz, P., & Rizvi, S. S. (2023). Hydrolysis of lactose in milk: Current status and future products. *Food Reviews International, 39*(5), 2875–2894.

Shrestha, A. K., Howes, T., Adhikari, B. P., & Bhandari, B. R. (2007). Water sorption and glass transition properties of spray dried lactose hydrolysed skim milk powder. *LWT-Food Science and Technology, 40*(9), 1593–1600.

Singh, P., Rao, P. S., Sharma, V., & Arora, S. (2021). Physico-chemical aspects of lactose hydrolysed milk system along with detection and mitigation of maillard reaction products. *Trends in Food Science & Technology, 107*, 57–67.

Sokoti Naji, L., Jafarian, P., Kalbasi, A., & Yarmand, S. (2017). Comparison of different milk permeate concentration (mineralized and demineralized) effect on oily cake quality characteristics. *Journal of Food Research, 27*(3), 103–113.

Torres, D. P., Gonçalves, M. D. P. F., Teixeira, J. A., & Rodrigues, L. R. (2010). Galacto oligosaccharides: Production, properties, applications, and significance as prebiotics. *Comprehensive Reviews in Food Science and Food Safety, 9*(5), 438–454.

Torres, J. K. F., Stephani, R., Tavares, G. M., De Carvalho, A. F., Costa, R. G. B., de Almeida, C. E. R., & Perrone, Í. T. (2017). Technological aspects of lactose-hydrolyzed milk powder. *Food Research International, 101*, 45–53.

Tossavainen, O., & Kallioinen, H. (2007). Proteolytic changes in lactose hydrolysed UHT milks during storage. *Milchwissenschaft-Milk Science International, 62*(4), 410–414.

Troise, A. D., Bandini, E., De Donno, R., Meijer, G., Trezzi, M., & Fogliano, V. (2016). The quality of low lactose milk is affected by the side proteolytic activity of the lactase used in the production process. *Food Research International, 89*, 514–525.

Tuohy, K. M., Ziemer, C. J., Klinder, A., Knöbel, Y., Pool-Zobel, B. L., & Gibson, G. R. (2002). A human volunteer study to determine the prebiotic effects of lactulose powder on human colonic microbiota. *Microbial Ecology in Health and Disease, 14*(3), 165–173.

Tvorogova, A. A., Shobanova, T. V., Gurskiy, I. A., & Kazakova, N. V. (2023). The quality indicators of ice cream upon the enzymatic hydrolysis of lactose. *Food Systems, 6*(3), 308–316.

Tweedie, L. S., & MacBean, R. D. (1978). The effect of partial replacement of sucrose by hydrolysed whey lactose on the quality of canned peaches and pears. *Food Technology in Australia, 30*(4), 128–131.

US Patent US20070274955A1. *Novel galactooligosaccharide composition and the preparation thereof.* Retrieved from https://patents.google.com/patent/US20070274955A1/en

Vera, C., Guerrero, C., Aburto, C., Cordova, A., & Illanes, A. (2020). Conventional and non-conventional applications of β-galactosidases. *Biochimica et Biophysica Acta (BBA)-proteins and proteomics, 1868*(1), 140271.

von Rymon Lipinski, G. W. (2013). Sweeteners. In H. Zorn & P. Czermak (Eds.), *Biotechnology of food and feed additives. Advances in biochemical engineering/biotechnology* (Vol. 143, pp. 1–28). Springer.

Yamamoto, E., Watanabe, R., Ichimura, T., Ishida, T., & Kimura, K. (2021). Effect of lactose hydrolysis on the milk-fermenting properties of *Lactobacillus delbrueckii* ssp. *bulgaricus* 2038 and *Streptococcus thermophilus* 1131. *Journal of Dairy Science, 104*(2), 1454–1464.

Zhang, H., Tao, Y., He, Y., Pan, J., Yang, K., Shen, J., & Gao, C. (2020). Preparation of low-lactose milk powder by coupling membrane technology. *ACS Omega, 5*(15), 8543–8550.

Index

P. S. Rao et al. (eds.), *Lactose Hydrolysis in Dairy Products*, https://doi.org/10.1007/978-3-031-78207-7

L

M

N

P

R

T

U

Y

MIX
Papier aus verantwortungsvollen Quellen
Paper from responsible sources
FSC® C105338

If you have any concerns about our products,
you can contact us on
ProductSafety@springernature.com

In case Publisher is established outside the EU,
the EU authorized representative is:
Springer Nature Customer Service Center GmbH
Europaplatz 3, 69115 Heidelberg, Germany

Printed by Libri Plureos GmbH
in Hamburg, Germany